Birkhäuser

Compact Textbooks in Mathematics

This textbook series presents concise introductions to current topics in mathematics and mainly addresses advanced undergraduates and master students. The concept is to offer small books covering subject matter equivalent to 2- or 3-hour lectures or seminars which are also suitable for self-study. The books provide students and teachers with new perspectives and novel approaches. They may feature examples and exercises to illustrate key concepts and applications of the theoretical contents. The series also includes textbooks specifically speaking to the needs of students from other disciplines such as physics, computer science, engineering, life sciences, finance.

- **compact**: small books presenting the relevant knowledge
- **learning made easy**: examples and exercises illustrate the application of the contents
- **useful for lecturers**: each title can serve as basis and guideline for a semester course/lecture/seminar of 2-3 hours per week.

Johann Davidov

Vector Bundles and Connections

An Introduction

Johann Davidov
Institute of Mathematics and Informatics
Bulgarian Academy of Sciences
Sofia, Bulgaria

ISSN 2296-4568 ISSN 2296-455X (electronic)
Compact Textbooks in Mathematics
ISBN 978-3-032-07405-8 ISBN 978-3-032-07403-4 (eBook)
https://doi.org/10.1007/978-3-032-07403-4

Mathematics Subject Classification: 53-01, 53C05, 58A99, 32L05

This book is published under the imprint Birkhäuser, www.birkhauser-science.com by the registered company Springer Nature Switzerland AG
The registered company address is: Gewerbestrasse 11, 6330 Cham, Switzerland

Preface

This book is an extended version of the lecture notes on a course the author has given for several years at Sofia University. Its main topic is the theory of connections on vector bundles. The concept of a connection is one of the central notions in differential geometry and is used also in other branches of geometry like algebraic and symplectic geometry, as well as in theoretical physics. An important application discussed in the book is the introduction and study of the Chern, Pontrjagin, and Euler classes (with real coefficients) by means of the Chern-Weil homomorphism. It is also worth noting that the book presents some results that are scattered in the literature and sometimes not very easy to be found therein.

After the first couple of lectures, it became clear that the audience was mixed, consisting of students with and without sufficient background. To keep the interest of both groups, the live lectures had to be balanced in providing details. In this book, the balance is changed in favour of the (upper level) undergraduate or graduate beginners, the discussions and proofs being presented in greater detail in order to be easily understandable and instructive. Readers that do not need this can omit (part of) the details.

Chapter 1 of the book titled "Vector Bundles" contains material about vector bundles which is more or less standard. Many examples illustrating the main concepts are provided. As an example of a bundle morphism, the notion of an almost complex structure is considered, as well as the notion of integrability of such a structure. The del-bar operator on a holomorphic vector bundle is discussed in the last paragraph.

Chapter 2 "Connections on Vector Bundles" is the main one. It concerns the theory of connections on vector bundles with special attention to the curvature of a connection. Connections compatible with a metric, or with an almost complex structure are considered, including the canonical (Chern) connection on a Hermitian holomorphic vector bundle. The curvature of a holomorphic subbundle of such a bundle is studied. The horizontal space of a connection is defined and it is shown how this space in its own turn yields the connection. The parallel transport along a curve determined by a connection is also examined.

In Chap. 3 "Tautological (Universal) Bundles", complex projective spaces, complex Grassmann manifolds, and the tautological (universal) bundles are considered, including a computation of the holomorphic tangent and canonical bundles of a Grassmann manifold, as well as the curvature of the tautological and tautological quotient bundles. The space of holomorphic sections of tensor powers of the tautological bundle and its dual over a complex projective space are computed. The Plücker embedding of a complex Grassmann manifold is defined. This embedding is used in the computation of the canonical bundle of a Grassmann manifold.

Chapter 4 "Chern, Pontrjagin, and Euler Classes of Vector Bundles" deals, as the title suggests, with these characteristic classes introduced via the Chern-Weil homomorphism involving connections and their curvature. At the beginning of the chapter, a short overview of de Rham cohomology is provided. The holomorphic line bundle determined by a complex hypersurface of a complex manifold is constructed and used for computing the Chern classes of hypersurfaces. The Pontrjagin classes of a real vector bundle are defined by means of the Chern classes of the complexified bundle. The notion of an oriented vector bundle is considered, and the Euler class of such a bundle is defined via the Pfaffian of the curvature of a metric connection on the bundle. The chapter ends with a brief discussion on a generalized version of the classical Gauss-Bonnet theorem (stated without proof).

Many exercise problems are given within each chapter. The initial idea for these lectures was inspired by Jean-Pierre Demailly's book *Complex Analytic and Differential Geometry*. In addition to the books listed in the bibliography, ideas and proofs suggested in personal conversations and public web discussions are also used.

I would like to thank Christian Yankov for proofreading the final manuscript and for suggesting some changes improving the text.

Sofia, Bulgaria Johann Davidov

Competing Interests The author has no competing interests to declare that are relevant to the content of this manuscript.

Contents

1 Vector Bundles

1.1 Definition and Examples of Vector Bundles

All manifolds considered in this course are assumed to be Hausdorff.

Let $\mathbb{K}$ be the field $\mathbb{R}$ or $\mathbb{C}$.

Definition 1.1.1 A smooth $\mathbb{K}$-vector bundle of rank r is a triple (E, M, π) of smooth manifolds E and M, and a smooth surjective map $\pi : E \to M$ such that:

(1) for every point $x \in M$, the set $E_x = \pi^{-1}(x)$ is a vector space over the field $\mathbb{K}$.

(2) for every point $x \in M$, there is an open neighbourhood U of the point x and a diffeomorphism

$$\varphi_U : \pi^{-1}(U) \to U \times \mathbb{K}^r,$$

whose restriction to every vector space E_y, $y \in U$, is an isomorphism onto the vector space $\{y\} \times \mathbb{K}^r$.

If in this definition $\mathbb{K} = \mathbb{C}$, E and M are complex manifolds, and the maps π, φ_U are holomorphic, we obtain the notion of a holomorphic vector bundle.

The condition that φ_U maps E_y onto $\{y\} \times \mathbb{K}^r$ may be stated as the requirement that the following diagramme is commutative:

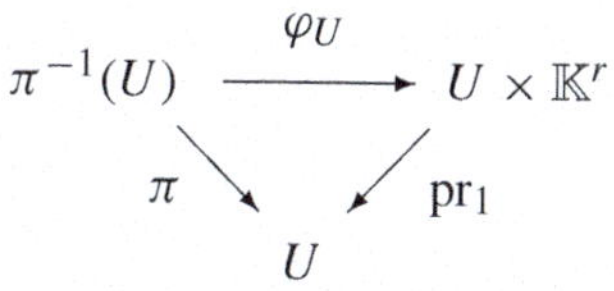

$$\begin{array}{ccc} \pi^{-1}(U) & \xrightarrow{\varphi_U} & U \times \mathbb{K}^r \\ & {\scriptstyle \pi} \searrow \quad \swarrow {\scriptstyle \mathrm{pr}_1} & \\ & U & \end{array}$$

J. Davidov, *Vector Bundles and Connections*, Compact Textbooks in Mathematics,
https://doi.org/10.1007/978-3-032-07403-4_1

In this diagramme, pr_1 is the natural projection of $U \times \mathbb{K}^r$ onto the first factor. The diffeomorphism φ_U is of the form $\varphi_U = (\pi, \Phi_U)$, where $\Phi_U : \pi^{-1}(U) \to \mathbb{K}^r$ is a smooth map which, for every $y \in U$, is an isomorphism of the vector space E_y onto the space $\mathbb{K}^r$.

It follows from (2) that the map π is open (i.e., sends an open set onto an open set). It follows also that π is a submersion (the rank of the differential π_* is equal to dim M at every point of E). Moreover, $dim\, E = dim\, M + r$. Since π is a submersion, every set E_x endowed with the topology induced from that of E possesses the structure of a (closed) submanifold of E of dimension $dim\, E - dim\, M = r$ (if necessary, see Warner [24], Chap. 1, Theorem 1.38 or Brickell and Clark [3], Chap. 6, Proposition 6.2.1). The restriction of the diffeomorphism φ_U to the submanifold E_x is a diffeomorphism of E_x onto $\{x\} \times \mathbb{K}^r$. This restriction is a linear isomorphism as well, hence the smooth structure of E_x coincides with that determined by the vector space structure of E_x.

The manifold E is called the **total space** of the bundle, M its **base**, and π the **projection**. Often, the manifold E itself is called a bundle over M. The set E_x is said to be the **fibre** of the bundle at the point x, and the diffeomorphism φ_U a **trivialization** of E over U. Such a neighbourhood U is called **trivializing**. The family $\{U, \varphi_U\}$ is said to be an **atlas** of the bundle E.

It follows from (2) that $\dim_{\mathbb{K}} E_x = r$ for every $x \in M$. Bundles of rank $r = 1$ are called **line bundles**.

Problem 1.1.1 *If φ_U is a trivialization and $V \subset U$ is an open set, $\varphi_V = \varphi_U|\pi^{-1}(V)$ is also a trivialization.*

Example 1.1.1 Let $\pi : M \times \mathbb{K}^r \to M$ be the natural projection $\pi(x, v) = x$. Then $M \times \mathbb{K}^r$ is a vector bundle over M of rank r called **trivial**.

Condition (2) means that every vector bundle locally, up to a diffeomeorphism, looks like the trivial bundle.

Example 1.1.2 Let $\pi : E \to M$ be a smooth vector bundle over a manifold M of rank r, and let N be a k-dimensional submanifold of M. Since π is a submersion, the set $F = \pi^{-1}(N)$ admits the structure of a submanifold of E of dimension $dim\, E - (dim\, M - dim\, N) = r + k$ (Warner [24], Chap. 1, Theorem 1.39 or Narasimhan [20], Chap. 2, § 2.16, Proposition 2.16.3). Moreover, if the topology of N is induced from that of M, the topology of F is induced from the topology of E. Set $\pi_F = \pi|F$. Then $\pi_F : F \to N$ is a smooth vector bundle of rank r. Indeed, if $\varphi_U : \pi^{-1}(U) \to U \times \mathbb{K}^r$ is a trivialization of E, the set $V = U \cap N$ is open in N, and the restriction of φ_U to $\pi_F^{-1}(V) = \pi^{-1}(U) \cap F$ is a diffeomorphism of $\pi_F^{-1}(V)$ onto $V \times \mathbb{K}^r$ (this restriction is smooth and bijective, and moreover is a local diffeomorphism since it is of rank $dim\, F = k + r$). The bundle F is called the **restriction of E over N** and is usually denoted by $E|N$.

Remark 1.1.1 Let N be a submanifold of a manifold M. The topology on N induced from that of M is contained in the topology of the manifold N since the inclusion map $\iota : N \hookrightarrow M$ is continuous. These two topologies on N may not coincide. For example, let $N = \{(\sin 2s, \sin s) : 0 \leq s \leq 2\pi\}$ be the figure eight along the y-axis. The map $s \to (\sin 2s, \sin s)$ bijectively maps the open interval $(0, 2\pi)$ onto N, hence N possesses a (uniquely determined) smooth structure for which this map provides a coordinate system. With this structure, N is a submanifold of $\mathbb{R}^2$. The manifold N is homeomorphic to an open interval, while the set N with the induced topology is a compact space. Therefore, the two topologies on N do not coincide.

A submanifold whose manifold topology coincides with the induced one is called **regular** or **embedded**. Open submanifolds are apparently embedded. Every compact submanifold N of a (Hausdorff) manifold M is embedded (if N' is the set N endowed with the induced topology, $Id : N \to N'$ is a bijective and continuous map of a compact space onto a Hausdorff space, hence a homeomorphism). As mentioned, the pre-image of an embedded submanifold (in particular, a point) under a submersion is also an embedded submanifold.

An important property of embedded submanifolds is the following: Let $f : M_1 \to M$ be a smooth map whose image lies in an embedded submanifold N of M. Then the map $g : M_1 \to N$ induced by f (i.e., $g(x) = f(x) \in N$, $x \in M_1$) is smooth (Warner [24], Chap. 1, Theorem 1.32 or Brickell and Clark [3], Chap. 5, Proposition 5.4.3). For non-embedded submanifolds, this statement is, generally speaking, not true (for a counterexample see Brickell and Clark [3], Chap. 5, Example 5.4.2). Using the statement above, it is easy to show that if N is an embedded submanifold of M of dimension k, then any smooth structure on the set N turning it into a submanifold N' of M of the same dimension k coincides with the given smooth structure of N (i.e., the identity map $Id : N' \to N$ is a diffeomorphism); the condition that N' is of the same dimension as N cannot be omitted (Brickell and Clark [3], Chap. 5, Example 5.4.4).

Example 1.1.3 The tangent bundle of the unit sphere.

Let $S^n = \{x = (x_1, \ldots, x_{n+1}) \in \mathbb{R}^{n+1} : |x| = 1\}$ be the unit n-sphere. For $x \in S^n$, denote by $T_x S^n$ the tangent space of S^n at the point x:

$$T_x S^n = \{v \in \mathbb{R}^{n+1} : v \perp x\}.$$

Set $TS^n = \{(x, v) \in S^n \times \mathbb{R}^{n+1} : v \perp x\}$, i.e., TS^n is the disjoint union $\dot{\bigcup}_x T_x S^n$ $(= \bigcup_{x \in S^n} (\{x\} \times T_x S^n))$. The set TS^n is determined by the equation $\sum_{i=1}^{n+1} v_i x_i = 0$, hence it is a smooth closed embedded hypersurface in $S^n \times \mathbb{R}^{n+1}$. Let $\pi : TS^n \to S^n$ be the map $(x, v) \to x$. Obviously $\pi^{-1}(x) = T_x S^n$. The sets $U_k = \{(x_1, \ldots, x_{n+1}) \in S^n : x_k \neq 0\}$, $1 \leq k \leq n+1$, constitute an open covering of S^n. If $(x, v) \in \pi^{-1}(U_k)$, then we have $x_1 v_1 + \cdots + x_k v_k + \cdots + x_{n+1} v_{n+1} = 0$ and $x_k \neq 0$, which implies $v_k = -\dfrac{1}{x_k} \sum_{i \neq k} x_i v_i$. Hence, the map $\varphi_k : \pi^{-1}(U_k) \to U_k \times \mathbb{R}^n$ defined by

$$(x, v) \to (x, v_1, \ldots, v_{k-1}, v_{k+1}, \ldots, v_n)$$

is a diffeomorphism. Clearly, the restriction of φ_k to $\pi^{-1}(x)$, $x \in U_k$, is an isomorphisms of $\pi^{-1}(x) = T_x S^n$ onto $\mathbb{R}^n$. Therefore $\pi : TS^n \to S^n$ is a smooth vector bundle over S^n of rank n.

Example 1.1.4 The normal bundle of the unit sphere.
The normal bundle of the unit sphere S^n is defined as

$$NS^n = \{(x, v) \in S^n \times \mathbb{R}^{n+1} : v \perp T_x S^n\}.$$

Set $U_k = \{x \in S^n : x_k \neq 0\}$. At every point $x = (x_1, \dots, x_{n+1}) \in U_k$, the vectors

$$X_1 = (\underbrace{-x_k, 0, \dots, 0, x_1}_{k}, 0, \dots, 0),\ X_2 = (0, -x_k, 0, \dots, 0, x_2, 0, \dots, 0),\ \dots,$$
$$X_n = (0, \dots, 0, x_{n+1}, 0, \dots, 0, -x_k)$$

constitute a basis of $T_x S^n$. Hence, in $U_k \times \mathbb{R}^{n+1}$, the set NS^n is given by the equation $-v_i x_k + v_k x_i = 0$, $i = 1, \dots, k-1, k+1, \dots, n+1$. Therefore NS^n is a closed $(n+1)$-dimensional embedded submanifold of $S^n \times \mathbb{R}^{n+1}$. Clearly, $NS^n = \{(x, v) \in S^n \times \mathbb{R}^{n+1} : v = tx$ for some $t \in \mathbb{R}\}$. The projection map $\pi : NS^n \to S^n$ is defined by $(x, v) \to x$. Let $\varphi_k : \pi^{-1}(U_k) \to U_k \times \mathbb{R}$ be the map

$$\varphi_k(x, v) = (x, \frac{v_k}{x_k}),\ k = 1, \dots, n+1.$$

The maps φ_k are local trivializations. Thus, $\pi : NS^n \to S^n$ is a line bundle, the normal bundle of S^n.

Example 1.1.5 Möbius strip.
Let D be the strip $D = \{(x, y) \in \mathbb{R}^2 : 0 \leq x \leq 1\}$. Identifying the points $(0, y)$ and $(1, -y)$ on the boundary of D, we obtain the Möbius strip with infinite width. Denote this strip by E. Thus, E is the quotient space of D under the equivalence relation defined by $(x, y) \sim (x', y')$ if $x = x',\ y = y'$ or $x = 0,\ x' = 1,\ y' = -y$ or $x = 1,\ x' = 0,\ y = -y'$. We denote the equivalence class of a point $(x, y) \in D$ by $[x, y]$. Set

$$U_1 = \{[x, y] : 0 < x < 1\},\quad U_1' = \{[x, y] : 0 \leq x < \frac{1}{2}\},\quad U_2' = \{[x, y] : \frac{1}{2} < x \leq 1\},$$

$$U_2 = U_1' \cup U_2'.$$

Obviously, the sets $U_1,\ U_1',\ U_2'$ form an open covering of E with respect to the quotient topology. Define maps $\varphi_1 : U_1 \to (0, 1)$ and $\varphi_2 : U_2 \to (-\frac{1}{2}, \frac{1}{2})$ as follows:

$$\varphi_1([x, y]) = (x, y),\quad \varphi_2([x, y]) = \begin{cases} (x, y) & \text{if } [x, y] \in U_1' \\ (x-1, -y) & \text{if } [x, y] \in U_2' \end{cases}$$

The map φ_2 is well-defined since $U_1' \cap U_2' = \{[0, y] = [1, -y] : y \in \mathbb{R}\}$. The maps φ_1 and φ_2 are homeomorphisms. Moreover, φ_2 sends U_1' onto $[0, \frac{1}{2}] \times \mathbb{R}$ and U_2' onto $[-\frac{1}{2}, 0] \times \mathbb{R}$. This implies that E is a Hausdorff space. Moreover

$$\varphi_1 \circ \varphi_2^{-1}(x, y) = \begin{cases} (x, y) & \text{if } 0 < x < \frac{1}{2} \\ (x + 1, -y) & \text{if } \frac{1}{2} < x < 1 \end{cases}$$

Hence, $\varphi_1 \circ \varphi_2^{-1}$ is a diffeomorphism. This shows that E has the structure of a smooth manifold with atlas $\{(U_1, \varphi_1), (U_2, \varphi_2)\}$.

Let S^1 be the unit circle obtained from the interval $[0, 1]$ by identifying the end points 0 and 1 (the map $[0, 1] \ni x \to e^{2\pi i x}$ yields a homeomorphism of the quotient space of $[0, 1]$ with S^1); if $x \in [0, 1]$, we denote its corresponding point on S^1 by $[x]$. Let $\pi : E \to S^1$ be the map $\pi([x, y]) = [x]$. For every $\xi = [x] \in S^1$,

$$\pi^{-1}(\xi) = \{[x, y] : y \in \mathbb{R}\},$$

where the set on the right-hand side does not depend on the choice of a representative x of the point ξ on S^1. Set

$$\theta_\xi([x', y']) = \begin{cases} y' & \text{if } 0 \le x' < 1 \\ -y' & \text{if } x' = 1 \end{cases}$$

Thus, we obtain a well-defined map θ_ξ of $\pi^{-1}(\xi)$ into $\mathbb{R}$. This map is bijective:

$$\theta_\xi^{-1}(y) = \begin{cases} [x, y] & \text{if } 0 \le x < 1 \\ [1, -y] & \text{if } x = 1 \end{cases}$$

Therefore, using the map θ_ξ, we can define on $\pi^{-1}(\xi)$ the structure of a one-dimensional real vector space for which θ_ξ is an isomorphism.

The sets

$$V_1 = \{[x] \in S^1 : 0 < x < 1\}, \quad V_1' = \{[x] \in S^1 : 0 \le x < \frac{1}{2}\},$$

$$V_2' = \{[x] \in S^1 : \frac{1}{2} < x \le 1\}$$

form an open covering of S^1 such that $V_1' \cap V_2' = [0] = [1]$. Clearly, $\pi^{-1}(V_1) = U_1$, $\pi^{-1}(V_1') = U_1'$, $\pi^{-1}(V_2') = U_2'$. Let $V_2 = V_1' \cup V_2'$. Define maps $\psi_1 : \pi^{-1}(U_1) \to V_1 \times \mathbb{R}$ and $\psi_2 : \pi^{-1}(U_2) \to V_2 \times \mathbb{R}$ setting

$$\psi_1([x, y]) = ([x], y)$$

$$\psi_2([x, y]) = \begin{cases} ([x], y) & \text{if } [x, y] \in U_1' \\ ([x], -y) & \text{if } [x, y] \in U_2' \end{cases}$$

It is not hard to show that the maps ψ_1 and ψ_2 are diffeomorphisms and $\psi_k | \pi^{-1}(\xi) = \theta_\xi$, $k = 1, 2$. Therefore $\pi : E \to S^1$ is a smooth line bundle.

Vector bundles are often constructed by means of the following scheme.

Proposition 1.1.1 *Let M be a smooth manifold, E a set, and $\pi : E \to M$ a surjective map. Suppose:*

(1) For every $x \in M$, $E_x = \pi^{-1}(x)$ is a $\mathbb{K}$-vector space.

(2) For every $x \in M$, there exist an open neighbourhood U of the point x and a bijective map

$$\varphi_U : \pi^{-1}(U) \to U \times \mathbb{K}^r,$$

which maps every vector space E_y, $y \in U$, isomorphically onto the vector space $\{y\} \times \mathbb{K}^r$.

(3) The maps

$$\varphi_U \circ \varphi_V^{-1} : (U \cap V) \times \mathbb{K}^r \to (U \cap V) \times \mathbb{K}^r$$

are smooth.

Then there exists a unique smooth structure on E such that $\pi : E \to M$ is a smooth vector bundle over M and the maps φ_U are its trivializations.

Proof Using the maps φ_U, we transfer the topology of $U \times \mathbb{K}^r$ onto $\pi^{-1}(U)$ and consider the disjoint union topology on E determined by the topologies of the sets $\pi^{-1}(U)$ (these sets cover E). Thus, a subset A of E is open if the set $\varphi_U(A \cap \pi^{-1}(U))$ is open in $U \times \mathbb{K}^r$ for every U. Apparently, all sets $\pi^{-1}(U)$ are open in this topology of E and the maps φ_U are homeomorphisms (the topology of E which we defined is the smallest topology with this property). For any open subset Ω of M, we have $\varphi_U(\pi^{-1}(\Omega) \cap \pi^{-1}(U)) = (\Omega \cap U) \times \mathbb{K}^r$, thus the set $\pi^{-1}(\Omega)$ is open in E. Hence, the map $\pi : E \to M$ is continuous. Note also that the topology of E is Hausdorff. Indeed, let $a, b \in E$, and set $x = \pi(a)$, $y = \pi(b)$. If $x \neq y$, then there exist disjoint neighbourhoods Ω_x and Ω_y of x and y, respectively. In this case, $\pi^{-1}(\Omega_x)$ and $\pi^{-1}(\Omega_y)$ are disjoint neighbourhoods of a and b. If $x = y$, the points a and b lie in one of the open sets $\pi^{-1}(U)$ each of which is homeomorphic to a Hausdorff space. Hence, if the points a and b are distinct, they possess disjoint neighbourhoods.

Now, using the maps φ_U, we transfer the charts (the smooth structure) of $U \times \mathbb{K}^r$ onto $\pi^{-1}(U)$, and for each chart (Q, ψ) of $U \times \mathbb{K}^r$, we set $\widetilde{Q} = \varphi_U^{-1}(Q)$ and $\widetilde{\psi} = \psi \circ \varphi_U$. Each set $\widetilde{Q}$ is open in E, and the map $\widetilde{\psi}$ is a homeomorphism of $\widetilde{Q}$ onto the open subset $\psi(Q)$ of $\mathbb{R}^n \times \mathbb{K}^r$. If (S, θ) is a chart of $V \times \mathbb{K}^r$, then it follows from the third assumption of the proposition that $\widetilde{\psi} \circ \widetilde{\theta}^{-1}$ is a diffeomorphism. Indeed, the set $\widetilde{\theta}(\widetilde{S} \cap \widetilde{Q})$ is open since $\widetilde{\theta}(\widetilde{S} \cap \widetilde{Q}) = \theta(S \cap (\varphi_V \circ \varphi_U^{-1})(Q))$. Furthermore, on $\widetilde{\theta}(\widetilde{S} \cap \widetilde{Q})$, we have $\widetilde{\psi} \circ \widetilde{\theta}^{-1} = \psi \circ (\varphi_U \circ \varphi_V^{-1}) \circ \theta^{-1}$. This shows that $\widetilde{\psi} \circ \widetilde{\theta}^{-1}$ is a diffeomorphism. Hence, E is a smooth manifold with an atlas composed of the charts $(\widetilde{Q}, \widetilde{\psi})$. With this smooth structure on E, $\pi : E \to M$ is a smooth vector bundle and the maps φ_U are its trivializations.

For any smooth structure on E such that $\pi : E \to M$ is a smooth bundle with trivializations φ_U, the maps φ_U are diffeomorphisms, i.e., the smooth structure on $\pi(U)$ is, in fact, the smooth structure on $U \times \mathbb{K}^r$ transferred to $\pi(U)$ by means of the maps φ_U. This proves the uniqueness of the structure on E with the desired properties. □

Example 1.1.6 The tangent bundle of a smooth manifold.

Let M be a n-dimensional smooth manifold. The tangent space T_pM of M at any point $p \in M$ is a real n-dimensional vector space. If $(U, x_1, \dots, x_n)$ is a chart of M, the tangent vectors $\frac{\partial}{\partial x_i}(p)$, $i = 1, \dots, n$, form a basis of T_pM, $p \in U$: every vector $v \in T_pM$ can be uniquely written as $v = \sum_{i=1}^n v_i \frac{\partial}{\partial x_i}(p)$ where $v_i = v(x_i) = dx_i(v)$. Let $TM = \dot{\bigcup\limits_{p \in M}} T_pM$ be the disjoint union of the tangent spaces of M. Let $\pi : TM \to M$ be the map sending every vector of T_pM to the point p. Denote by φ_U the map

$$\varphi_U : \pi^{-1}(U) \ni v \to (\pi(v), dx_1(v), \dots, dx_n(v)).$$

It is easy to check that the assumptions of Proposition 1.1.1 are satisfied. Hence, TM is a smooth vector bundle over M, the tangent bundle of the manifold M. By the very definition of the smooth structure on TM, it is clear that

$$(x_1 \circ \pi, \dots, x_n \circ \pi, dx_1, \dots, dx_n)$$

is a coordinate system of TM on $\pi^{-1}(U)$.

Problem 1.1.2 *If $(V, y_1, \dots, y_n)$ is a chart of M with $V \cap U \neq \emptyset$, find an explicit formula for $\varphi_U \circ \varphi_V^{-1}$.*

Example 1.1.7 The holomorphic tangent bundle of a complex manifold.

Let M be a n-dimensional complex manifold. For $p \in M$, denote by T_p^hM the holomorphic tangent space of M at the point p, the complex space of derivations of local holomorphic functions at p. Similar to the case of a smooth manifold, the disjoint union $T^hM = \dot{\bigcup\limits_{p \in M}} T_p^hM$ can be endowed with the structure of a holomorphic vector bundle of rank n.

1.2 Morphisms of Vector Bundles

Intuitively, we can say that two bundles are "indistinguishable", i.e., isomorphic when there is a bijective map preserving both the smooth structure of the total space and the vector space structure on the fibres. Therefore the following definition is quite natural.

Definition 1.2.1 A morphism (or a bundle map) of a $\mathbb{K}$-vector bundle $\pi' : E' \to M$ into another $\mathbb{K}$-vector bundle $\pi'' : E'' \to M$ is a smooth map $f : E' \to E''$ sending every fibre E'_x, $x \in M$, into the fibre E''_x, i.e., $\pi'' \circ f = \pi'$, such that the restriction $f|E'_x : E'_x \to E''_x$ is a $\mathbb{K}$-linear map.

A morphism $f : E' \to E''$ is called an isomorphism if there exists a morphism $g : E'' \to E'$ such that $f \circ g = Id$ and $g \circ f = Id$.

Using the notion of a morphism, we may say that condition (2) in the definition of a vector bundle means that every point $x \in M$ admits a neighbourhood U such that the bundle E over U (i.e., the restriction $E\big|U = \pi^{-1}(U)$) is isomorphic to the trivial bundle $U \times \mathbb{K}^r$.

Example 1.2.1 Let N be a submanifold of a manifold M, and let $\iota : N \hookrightarrow M$ be the inclusion map. For every point $p \in N$, there exist local coordinates $x_1, \ldots, x_n$ of M on a neighbourhood of p such that $y_1 = x_1 \circ \iota, \ldots, y_k = x_k \circ \iota, k = dim\, N$, are local coordinates of N in a neighbourhood of p (see, for example, Warner [24], Chap. 1, Sect. 1.30, Corollary (f)). In these coordinates, $\iota_*(\frac{\partial}{\partial y_l}) = \frac{\partial}{\partial x_l}, l = 1, \ldots, k$. Hence, the map $\iota_* : TN \to TM$ is smooth. Moreover, it is linear on the fibres. Hence, the map i_* is a morphism of bundles. Clearly, it is injective.

Example 1.2.2 Let NS^n be the normal bundle of the sphere S^n. For every point $x \in S^n$, (x, x) is a basis of the fibre $N_x S^n$. Thus, for every $(x, v) \in NS^n$, the vector v is can be uniquely written as $v = tx$, $t \in \mathbb{R}$. The map $(x, v) \to (x, t)$ is an isomorphism of NS^n onto the trivial bundle $S^n \times \mathbb{R}$.

Example 1.2.3 Every complex manifold M possesses a natural smooth structure such that if $z_1 = x_1 + iy_1, \ldots, z_n = x_n + iy_n, x_k = Re\, z_k, y_k = Im\, z_k$, are complex coordinates, $x_1, \ldots x_n, y_1, \ldots, y_n$ are smooth coordinates. For the moment, denote by M^{re} the smooth manifold so described. Let TM^{re} be the tangent bundle of this manifold. Consider the holomorphic tangent bundle $T^h M$ of the complex manifold M as a smooth real vector bundle. Denote, for the moment, this real bundle by $(T^h M)^{re}$. The smooth bundles TM^{re} and $(T^h M)^{re}$ are canonically isomorphic via the map defined in the following way. Let $X \in T_p M^{re}$. For every holomorphic function $f = u + iv$ in a neighbourhood of the point p, set $\widehat{X}(f) = X(u) + iX(v)$. Then $\widehat{X} \in T^h_p M$ and the correspondence $X \to \widehat{X}$ is $\mathbb{R}$-linear. If $z_k = x_k + iy_k$, $k = 1, \ldots, n$, is a complex coordinate system of M, then, under this correspondence, $\frac{\partial}{\partial x_k} \to \frac{\partial}{\partial z_k}$ and $\frac{\partial}{\partial y_k} \to i\frac{\partial}{\partial z_k}$ since, for every holomorphic function f, we have $\frac{\partial f}{\partial x_k} = \frac{\partial f}{\partial z_k}$ and $\frac{\partial f}{\partial y_k} = i\frac{\partial f}{\partial z_k}$. This shows that the map $X \to \widehat{X}$ is an isomorphism of TM^{re} onto $(T^h M)^{re}$.

Let $f : E' \to E''$ be a map sending a fibre into a fibre. Then, if φ' and φ'' are trivializations of E' and E'' over an open set U, we can consider the map $\varphi'' \circ f \circ (\varphi')^{-1} : U \times \mathbb{K}^r \to U \times \mathbb{K}^s$. This map is of the form $(x, \xi) \to (x, \psi(x, \xi))$. For every fixed $x \in U$, let $\psi_x : \mathbb{K}^r \to \mathbb{K}^s$ be the map $\xi \to \psi(x, \xi)$. Then

$$\psi_x = (\varphi''|E_x'') \circ (f|E_x') \circ (\varphi')^{-1}|(\{x\} \times \mathbb{K}^r).$$

Hence, the map $f|E_x' : E_x' \to E_x''$ is linear exactly when ψ_x is a linear map.

Suppose that all maps ψ_x are linear. The following simple statement gives a relation between the smoothness of the map $\psi : U \times \mathbb{K}^r \to \mathbb{K}^s$ and the map $x \to \psi_x$ of U into the vector space $Hom(\mathbb{K}^r, \mathbb{K}^s)$ of linear maps $\mathbb{K}^r \to \mathbb{K}^s$.

Lemma 1.2.1 *The map $\psi : U \times \mathbb{K}^r \to \mathbb{K}^s$ is smooth if and only if the map $U \to Hom(\mathbb{K}^r, \mathbb{K}^s)$, $x \to \psi_x$, is smooth.*

Proof The map $x \to \psi_x$ is smooth exactly when the entries of the matrix of the linear map ψ_x with respect to fixed bases of $\mathbb{K}^r$ and $\mathbb{K}^s$ are smooth functions of x. The lemma follows immediately from this fact since $\psi(x, \xi)$ is the action of this matrix on the vector ξ. □

Note that if the map $x \to \psi_x$ is smooth and every map ψ_x is invertible, i.e., is an isomorphism, then, according to the standard formula for computing the inverse matrix, the map $x \to \psi_x^{-1}$ is smooth. We use this remark and Lemma 1.2.1 in the proof of the following statement.

Proposition 1.2.1 *A morphism $f : E' \to E''$ of vector bundles is an isomorphism if and only if, for every $x \in M$, the map $f|E_x' : E_x' \to E_x''$ is an isomorphism between the vector space E_x' and E_x''.*

Proof Necessity follows from the obvious fact that if $f^{-1} : E'' \to E'$ is the inverse morphism of f, the restriction $f^{-1}|E_x''$ is the inverse map of $f|E_x'$.

Sufficiency. Define a map $g : E'' \to E'$ setting $g|E_x'' = (f|E_x')^{-1}$. Obviously, g is the inverse of the map f, sends a fibre into a fibre, and is an isomorphism on the fibres. Thus, to show that g is a morphism, it remains to prove that g is a smooth map. As above, let φ' and φ'' be trivializations of E' and E'' on an open set U. Put $u = \varphi'' \circ f \circ (\varphi')^{-1}$, $v = \varphi' \circ g \circ (\varphi'')^{-1}$. Let $u(x, \xi) = (x, \psi(x, \xi))$, and let ψ_x be the map defined by $\psi_x(\xi) = \psi(x, \xi)$. Since $f|E_x'$ is an isomorphism, ψ_x is also an isomorphism and $v(x, \xi) = (x, \psi_x^{-1}(\xi))$. By Lemma 1.2.1, the map $x \to \psi_x$ of U into $Hom(\mathbb{K}^r, \mathbb{K}^r)$ is smooth. Then the map $x \to \psi_x^{-1}$ is also smooth and Lemma 1.2.1 implies that the map $(x, \xi) \to \psi_x^{-1}(\xi)$ is smooth. Hence, the map v is smooth. It follows that g is a smooth map. □

1.3 Transition Functions of a Vector Bundle

Let $\pi : E \to M$ be a smooth vector bundle, and let $\varphi_U = (\pi, \Phi_U)$, $\varphi_V = (\pi, \Phi_V)$ be two trivializations of E. For $(x, \xi) \in V \times \mathbb{K}^r$, we have $\varphi_V^{-1}(x, \xi) =$

$(\Phi_V|E_x)^{-1}(\xi)$. Therefore, if $U \cap V \neq \emptyset$, the map

$$\varphi_U \circ \varphi_V^{-1} : (U \cap V) \times \mathbb{K}^r \to (U \cap V) \times \mathbb{K}^r$$

is of the form

$$\varphi_U \circ \varphi_V^{-1}(x, \xi) = (x, (\Phi_U|E_x) \circ (\Phi_V|E_x)^{-1}(\xi)); \;\; x \in U \cap V, \;\; \xi \in \mathbb{K}^r.$$

For brevity, set $g_{UV}(x, \xi) = (\Phi_U|E_x) \circ (\Phi_V|E_x)^{-1}(\xi)$. Then the map

$$g_{UV}(x) : \xi \to g_{UV}(x, \xi)$$

is an automorphism of the vector space $\mathbb{K}^r$. Thus, we obtain a map

$$g_{UV} : U \cap V \to GL(\mathbb{K}^r) \tag{1.1}$$

of $U \cap V$ into the group $GL(\mathbb{K}^r)$ of invertible linear transformations of $\mathbb{K}^r$. It follows from the identity $\varphi_U \circ \varphi_V^{-1}(x, \xi) = (x, g_{UV}(x, \xi))$ that the map $g_{UV}(x, \xi) : (U \cap V) \times \mathbb{K}^r \to \mathbb{K}^r$ is smooth. Then, by Lemma 1.2.1, g_{UV}, considered as a map from $U \cap V$ into the space $Hom(\mathbb{K}^r, \mathbb{K}^r)$, is smooth. Since $GL(\mathbb{K}^r)$ is an open subset of $Hom(\mathbb{K}^r, \mathbb{K}^r)$, we conclude that the map (1.1) is smooth.

The map g_{UV} is called the **transition function** of E determined by the trivializations φ_U and φ_V. Clearly, the transition functions satisfy the identity

$$g_{UV}(x) \circ g_{VW}(x) \circ g_{WU}(x) = Id \;\; \text{for } x \in U \cap V \cap W.$$

Since $g_{UV}(x)$ are invertible maps, it follows that

$$g_{UV}(x) = Id \;\; \text{and} \;\; g_{VU}(x) = (g_{UV}(x))^{-1}$$

(which also follows from the very definition of the transition functions).

Problem 1.3.1 *Let M be a smooth manifold. Find the transition functions of the tangent bundle TM with respect to the trivializations defined by means of local coordinates of M.*

For every $e \in E$, consider the images of e in the sets $U \times \mathbb{K}^r$ under the maps φ_U. We can consider that these images "represent" e. If $(x, \xi) = \varphi_U(e) \in U \times \mathbb{K}^r$ and $(y, \eta) = \varphi_V(e) \in V \times \mathbb{K}^r$ are two "representations" of e, then $(x, \xi) = (y, g_{UV}(y).\eta)$. This relation between two "representations" shows that the bundle E can be obtained from the sets $U \times \mathbb{K}^r$ gluing any two sets $U \times \mathbb{K}^r$ and $V \times \mathbb{K}^r$ on their common part $(U \cap V) \times \mathbb{K}^r$ by means of the map g_{UV}.

Proposition 1.3.1 *Let M be a smooth manifold. Let $\{U_\alpha\}$ be an open cover of M and $g_{\alpha\beta} : U_\alpha \cap U_\beta \to GL(\mathbb{K}^r)$, $U_\alpha \cap U_\beta \neq 0$, smooth maps such that*

$$\begin{aligned} &g_{\alpha\alpha}(x) = Id \;\; \textit{for} \;\; x \in U_\alpha \\ &g_{\alpha\beta}(x) \circ g_{\beta\gamma}(x) \circ g_{\gamma\alpha}(x) = Id \;\; \textit{for} \;\; x \in U_\alpha \cap U_\beta \cap U_\gamma. \end{aligned} \tag{1.2}$$

Then there exists a unique, up to isomorphism, smooth $\mathbb{K}$-vector bundle $\pi : E \to M$ for which $g_{\alpha\beta}$ are transition functions.

Proof The notes preceding the proposition suggest the following construction. Set

$$\widetilde{E} = \dot{\bigcup_\alpha}(U_\alpha \times \mathbb{K}^r), \text{ disjoint union.}$$

Define the following equivalence relation on $\widetilde{E}$:

$$(\alpha, x, \xi) \sim (\beta, y, \eta) \text{ for } (x, \xi) \in U_\alpha \times \mathbb{K}^r \text{ and } (y, \eta) \in U_\beta \times \mathbb{K}^r$$
$$\text{exactly when } x = y \text{ and } \xi = g_{\alpha\beta}(y).\eta.$$

Now, set $E = \widetilde{E}/\sim$ and define a map $\pi : E \to M$ by $[(\alpha, x, \xi)] \to x$. Every element of $\pi^{-1}(U_\alpha)$ has a unique representative of the form (α, x, ξ) with $x \in U_\alpha$, $\xi \in \mathbb{K}^r$. Define a map $\varphi_\alpha : \pi^{-1}(U_\alpha) \to U_\alpha \times \mathbb{K}^r$ by $\varphi_\alpha([\alpha, x, \xi]) = (x, \xi)$. Then $\varphi_\alpha \circ \varphi_\beta^{-1}(x, \zeta) = (x, g_{\alpha\beta}(x).\zeta)$. It follows from Proposition 1.1.1 that $\pi : E \to M$ is a smooth vector bundle with transition functions $g_{\alpha\beta}$.

To prove the uniqueness, let $\pi' : E' \to M$ be also a vector bundle such that $g_{\alpha\beta}$ are transition functions with respect to some trivializations $\psi_\alpha : \pi^{-1}(U_\alpha) \to U_\alpha \times \mathbb{K}^r$. Then, for $e \in \pi^{-1}(U_\alpha)$, set

$$f(e) = \psi_\alpha^{-1} \circ \varphi_\alpha(e).$$

It is easy to see that $f(e)$ does not depend on the choice of α. Indeed, for $(x, \zeta) \in (U_\alpha \cap U_\beta) \times \mathbb{K}^r$,

$$\psi_\beta \circ \psi_\alpha^{-1} \circ \varphi_\alpha \circ \varphi_\beta^{-1}(x, \zeta) = (x, g_{\beta\alpha}(x) \circ g_{\alpha\beta}(x).\zeta) = (x, \zeta)$$

Hence, for $(x, \zeta) = \varphi_\beta(e)$ with $e \in \pi^{-1}(U_\alpha) \cap \pi^{-1}(U_\beta)$,

$$\psi_\beta \circ \psi_\alpha^{-1} \circ \varphi_\alpha(e) = \varphi_\beta(e).$$

Therefore $\psi_\alpha^{-1} \circ \varphi_\alpha(e) = \psi_\beta^{-1} \circ \varphi_\beta(e)$. Thus, we obtain a well-defined map $f : E \to E'$. This map is an isomorphism of vector bundles, its inverse map is given by $g(e') = \varphi_\alpha^{-1} \circ \psi_\alpha(e')$ for $e' \in \pi'^{-1}(U_\alpha)$. □

Remark 1.3.1 Identities (1.2) are called condition for $\{g_{\alpha\beta}\}$ to be a **cocycle**. They imply that $g_{\beta\alpha}(x) = g_{\alpha\beta}(x)^{-1}$. Identities (1.2) are equivalent to the identity $g_{\alpha\beta}(x) \circ g_{\beta\gamma}(x) = g_{\alpha\gamma}(x)$.

Remark 1.3.2 Let us endow $\widetilde{E}$ with the disjoint union topology: a subset of $\widetilde{E}$ is open if its intersection with every $U_\alpha \times \mathbb{K}^r$ is open in $U_\alpha \times \mathbb{K}^r$. Then the trivializations φ_α are homeomorphisms with respect to the quotient topology of $E = \widetilde{E}/\sim$. Hence, the topology of E coincides with its quotient topology.

We shall discuss a little further the problem of when transition functions corresponding to different coverings of M yield isomorphic bundles. To this end, we shall first deal with the local representations of morphisms between bundles.

Let $E_1 \to M$ and $E_2 \to M$ be vector bundles of rank r and s, and let $u : E_1 \to E_2$ be a morphism between them. Let $\{U_\alpha, \varphi_\alpha\}$ and $\{V_i, \psi_i\}$ be atlases of E_1 and E_2 with transition functions $f_{\alpha\beta}$ and g_{ij}. If $U_\alpha \cap V_i \neq \emptyset$, we can consider the composition

$$(U_\alpha \cap V_i) \times \mathbb{K}^r \xrightarrow{\varphi_\alpha^{-1}} E_1|U_\alpha \cap V_i \xrightarrow{u} E_2|U_\alpha \cap V_i \xrightarrow{\psi_i} (U_\alpha \cap V_i) \times \mathbb{K}^s$$

This composition is of the form

$$\psi_i \circ u \circ \varphi_\alpha^{-1}(x, \xi) = (x, u_{i\alpha}(x).\xi), \tag{1.3}$$

where $u_{i\alpha}(x) : \mathbb{K}^r \to \mathbb{K}^s$ is a linear map. Then, for every $x \in U_\alpha \cap U_\beta \cap V_i \cap V_j$,

$$u_{j\beta}(x) = g_{ji}(x) \circ u_{i\alpha}(x) \circ f_{\alpha\beta}(x). \tag{1.4}$$

Indeed, in view of (1.3), we have for $\xi \in \mathbb{K}^r$

$$\begin{aligned}(x, u_{j\beta}(x).\xi) &= \psi_j \circ u \circ \varphi_\alpha^{-1} \circ \varphi_\alpha \circ \varphi_\beta^{-1}(x, \xi)\\ &= \psi_j \circ u \circ \varphi_\alpha^{-1}(x, f_{\alpha\beta}(x).\xi)\\ &= \psi_j \circ \psi_i^{-1} \circ \psi_i \circ u \circ \varphi_\alpha^{-1}(x, f_{\alpha\beta}(x).\xi)\\ &= \psi_j \circ \psi_i^{-1}(x, u_{i\alpha}(x) \circ f_{\alpha\beta}(x).\xi)\\ &= (x, g_{ji}(x) \circ u_{i\alpha}(x) \circ f_{\alpha\beta}(x).\xi)\end{aligned}$$

This implies (1.4). Note that, by Lemma 1.2.1, the map $U_\alpha \cap V_i \to Hom(\mathbb{K}^r, \mathbb{K}^s)$, $x \to u_{i\alpha}(x)$ is smooth.

Conversely, suppose that for every two indices α and i for which $U_\alpha \cap V_i \neq \emptyset$ a smooth map $u_{i\alpha} : U_\alpha \cap V_i \to Hom(\mathbb{K}^r, \mathbb{K}^s)$ is given such that identity (1.4) is satisfied. Then we can define a map $u : E_1 \to E_2$ setting for $e \in E_1|U_\alpha \cap V_i$

$$u(e) = \psi_i^{-1}(x, u_{i\alpha}(x).\xi), \text{ where } (x, \xi) = \varphi_\alpha(e). \tag{1.5}$$

If $e \in E_1|U_\beta \cap V_j$, set $(y, \eta) = \varphi_\beta(e)$. Then $y = x = \pi(e)$ and $(x, \xi) = \varphi_\alpha(e) = \varphi_\alpha \circ \varphi_\beta^{-1}(y, \eta) = (y, f_{\alpha\beta}(y).\eta)$. Therefore, by (1.4),

$$\begin{aligned}&\psi_i^{-1}(x, u_{i\alpha}(x).\xi) = \psi_j^{-1} \circ \psi_j \circ \psi_i^{-1}(y, u_{i\alpha}(y) \circ f_{\alpha.\beta}(y).\eta)\\ &= \psi_j^{-1}(y, g_{ji}(y) \circ u_{i\alpha}(y) \circ f_{\alpha\beta}(y).\eta) = \psi_j^{-1}(y, u_{j\beta}(y).\eta).\end{aligned}$$

This shows that the map u is well-defined. It is clear from its definition that u is a morphism of bundles and that $\psi_i \circ u \circ \varphi_\alpha^{-1}(x, \xi) = (x, u_{i\alpha}(x).\xi)$.

We shall say that **the family** $\{u_{i\alpha}\}$ **represents the morphism** u.

Now, let $E_3 \to M$ be a vector bundle with an atlas $\{W_a, \theta_a\}$ and transition functions $\{h_{ab}\}$. Let $v : E_2 \to E_3$ be the morphism represented by a family $\{v_{ai}\}$. If the morphism $w = v \circ u : E_1 \to E_3$ is represented by $\{w_{a\alpha}\}$, then for every

$x \in U_\alpha \cap V_i \cap W_a$

$$w_{a\alpha}(x) = v_{ai}(x) \circ u_{i\alpha}(x).$$

Indeed,

$$\begin{aligned}(x, w_{a\alpha}(x).\xi) &= \theta_a \circ v \circ u \circ \varphi_\alpha^{-1}(x, \xi) \\ &= (\theta_a \circ v \circ \psi_i^{-1}) \circ (\psi_i \circ u \circ \varphi_\alpha^{-1}(x, \xi)) \\ &= \theta_a \circ v \circ \psi_i^{-1}(x, u_{i\alpha}(x).\xi) \;=\; (x, v_{ai}(x) \circ u_{i\alpha}(x).\xi)\end{aligned}$$

As a consequence, we see that if $u_{i\alpha}(x),\ x \in U_\alpha \cap V_i$, are isomorphisms, the morphism u defined via (1.5) is an isomorphism (this also follows from Proposition 1.2.1); the inverse isomorphism is represented by the maps $(u^{-1})_{\alpha i}(x) = u_{i\alpha}(x)^{-1}$.

In the case when the maps $u_{i\alpha}(x)$ are isomorphisms (equivalently, u is an isomorphism), identity (1.4) can be written as

$$g_{ij}(x) = u_{i\alpha}(x) \circ f_{\alpha\beta}(x) \circ u_{j\beta}(x)^{-1}, \quad x \in U_\alpha \cap U_\beta \cap V_i \cap V_j. \tag{1.6}$$

Thus, two sets of functions $f_{\alpha\beta} : U_\alpha \cap U_\beta \to GL(\mathbb{K}^r)$ and $g_{ij} : V_i \cap V_j \to GL(\mathbb{K}^r)$ satisfying the cocycle condition determine isomorphic bundles if and only if there exist smooth maps $u_{i\alpha} : U_\alpha \cap V_i \to GL(\mathbb{K}^r)$ satisfying identity (1.6).

Now, let $E = M \times \mathbb{K}^r$ be the trivial bundle. In this case, we can take the atlas $\{U_\alpha, \varphi_\alpha\}$ of E with $U_\alpha = M$, $\varphi_\alpha = Id$. For this atlas, $f_{\alpha\beta}(x) = Id$, and setting $u_i(x) = u_{i\alpha}(x)$, the isomorphism condition (1.6) for E and a bundle F takes the form

$$g_{ij}(x) = u_i(x) \circ u_j(x)^{-1}, \quad x \in V_i \cap V_j,$$

where $u_i : V_i \to GL(\mathbb{K}^r)$ are smooth maps. A cocycle g_{ij} that can be written in this form is called a **coboundary**. Thus, a vector bundle is isomorphic to the trivial one if and only if it admits transition functions whose cocycle is a coboundary.

1.4 Algebraic Operations on Vector Bundles

As a rule, the algebraic operations on vector spaces can be carried on vector bundles.

1. Let $\pi_E : E \to M$ and $\pi_F : F \to M$ be smooth $\mathbb{K}$-vector bundles over a manifold M. Set

$$E \oplus F = \dot{\bigcup_{x \in M}} (E_x \oplus F_x), \text{ disjoint union.}$$

Let $\pi : E \oplus F \to M$ be the map sending $E_x \oplus F_x$ to the point x. For every $x \in M$, there exist a neighbourhood U of x and local trivializations of E and F on U

$$\varphi_U^E = (\pi_E, \Phi_U^E) : \pi_E^{-1}(U) \to U \times \mathbb{K}^r, \quad \varphi_U^F = (\pi_F, \Phi_U^F) : \pi_F^{-1}(U) \to U \times \mathbb{K}^s.$$

Define a map

$$\varphi_U : \pi^{-1}(U) \to U \times (\mathbb{K}^r \oplus \mathbb{K}^s)$$

by

$$\varphi_U(a+b) = (x, \Phi_U^E(a) + \Phi_U^F(b))$$

for $a + b \in E_x \oplus F_x$. This map is bijective and $\mathbb{K}$-linear on the fibres $\pi^{-1}(x)$. Moreover,

$$\varphi_U \circ \varphi_V^{-1}(x, \xi + \eta) = (x, g_{UV}^E(x).\xi + g_{UV}^F(x).\eta)$$

for $x \in U \cap V$, $\xi \in \mathbb{K}^r$, $\eta \in \mathbb{K}^r$. In particular, it follows that the maps $\varphi_U \circ \varphi_V^{-1}$ are smooth. Hence, by Proposition 1.1.1, $E \oplus F$ admits the structure of a smooth $\mathbb{K}$-vector bundle over M with transition functions $g_{UV}^E(x) \oplus g_{UV}^F(x) \in GL(\mathbb{K}^r \oplus \mathbb{K}^s)$ or in block-matrix form

$$\begin{bmatrix} g_{UV}^E & 0 \\ 0 & g_{UV}^F \end{bmatrix}.$$

The following bundles can be defined in a similar way:

2. $E \otimes F$ with transition functions $g_{UV}^E(x) \otimes g_{UV}^F(x) \in GL(\mathbb{K}^r \otimes \mathbb{K}^s)$.

3. Set $E^* = \dot{\bigcup}_{x \in M} E_x^*$, the disjoint union of the dual spaces E_x^*. Define a map $\pi_{E^*} : E^* \to M$ by $\pi_{E^*}(E_x^*) = x$. Let $\varphi_U = (\pi^E, \Phi_U)$ be a trivialization of the bundle E. Denote by $\chi : \mathbb{K}^r \to (\mathbb{K}^r)^*$ the isomorphism sending the standard basis $e_1, \ldots, e_r$ of $\mathbb{K}^r$ to its dual basis $\eta_1, \ldots, \eta_r$. Let $\Psi_U : \pi_{E^*}^{-1}(U) = \dot{\bigcup}_{x \in U} E_x^* \to \mathbb{K}^r$ be the map whose restriction $\Psi|E_x^*$ is the isomorphism $E_x^* \to \mathbb{K}^r$ defined by $\Psi|E_x^* = ((\Phi_U|E_x)^* \circ \chi)^{-1}$ where $(\Phi_U|E_x)^* : (\mathbb{K}^r)^* \to E_x^*$ is the dual map of the linear map $\Phi|E_x : E_x \to \mathbb{K}^r$. Now, denote the map $(\pi_{E^*}, \Psi_U) : \pi_{E^*}^{-1}(U) \to U \times \mathbb{K}^r$ by ψ_U. Let $\varphi_V = (\pi^E, \Phi_V)$ be another trivialization of the bundle E and let ψ_V be the corresponding map $\psi_V = (\pi_{E^*}, \Psi_V) : \pi_{E^*}^{-1}(V) \to V \times \mathbb{K}^r$. Denote by g_{UV} the transition function of the bundle E determined by the trivializations φ_U and φ_V. For $x \in U \cap V$, let $[g_{ij}(x)]$ be the matrix of the transformation $g_{UV}(x) \in GL(\mathbb{K}^r)$ with respect to the standard basis $e_1, \ldots, e_r$ of $\mathbb{K}^r$. Thus, $(\Phi_U|E_x) \circ (\Phi_V|E_x)^{-1}(e_i) = \sum_{j=1}^r g_{ij}(x)e_j$, $i = 1, \ldots, r$, or

$$(\Phi_V|E_x)^{-1}(e_i) = \sum_{j=1}^r g_{ij}(x)(\Phi_U|E_x)^{-1}(e_j). \tag{1.7}$$

The basis $(\Psi_U|E_x^*)^{-1}(e_i) = ((\Phi_U|E_x)^*)^{-1}(\eta_i)$ of E_x^* is dual to the basis $(\Phi_U|E_x)^{-1}(e_j)$ of E_x. Similarly for the bases $(\Psi_V|E_x^*)^{-1}(e_i)$ and $(\Phi_V|E_x)^{-1}(e_j)$. It follows from (1.7) that if

$$(\Psi_V|E_x^*)^{-1}(e_i) = \sum_{j=1}^r g_{ij}^*(x)(\Psi_U|E_x^*)^{-1}(e_j),$$

then $[g_{ij}^*(x)] = ([g_{ij}(x)]^{-1})^t$, the transpose matrix of the inverse matrix $[g_{ij}(x)]^{-1}$. The functions g_{ij} are smooth on $U \cap V$. By the standard rule for finding the inverse matrix, the entries of the matrix $[g_{ij}(x)]^{-1}$ are also smooth. Hence, the map $\psi_U \circ \psi_V(x, \xi) = (x, ([g_{UV}(x)]^{-1})^t.\xi)$ is smooth; here "t" means the conjugate operator with respect to the standard metric of $\mathbb{K}^r$. By Proposition 1.1.1, $\pi_{E^*} : E^* \to M$ admits the structure of a vector bundle of rank r with transition functions $([g_{UV}(x)]^{-1})^t$.

4. $Hom(E, F) = \dot{\bigcup\limits_{x \in M}} Hom(E_x, F_x)$. The projection $\pi : Hom(E, F) \to M$ is the map sending $Hom(E_x, F_x)$ to the point x. For $f \in Hom(E_x, F_x)$ with $x \in U$

$$\varphi_U^F \circ f \circ (\varphi_U^E)^{-1}(x, \xi) = (x, \widehat{f_U}(x).\xi), \ \xi \in \mathbb{K}^r,$$

where $\widehat{f_U}(x) : \mathbb{K}^r \to \mathbb{K}^s$ is a linear map. Define a map $\varphi_U : \pi^{-1}(U) \to U \times Hom(\mathbb{K}^r, \mathbb{K}^s)$ by $\varphi_U(f) = (x, \widehat{f_U}(x)), \ x = \pi(f)$.

Let $x \in U \cap V$, $u \in Hom(\mathbb{K}^r, \mathbb{K}^s)$, and set $f = \varphi_V^{-1}(x, u)$. Then $\varphi_V^F \circ f \circ (\varphi_V^E)^{-1}(x, \xi) = (x, u(\xi))$ for $\xi \in \mathbb{K}^r$. Hence

$$\begin{aligned}(x, \widehat{f_U}(x).\xi) &= \varphi_U^F \circ (\varphi_V^F)^{-1} \circ (\varphi_V^F \circ f \circ (\varphi_V^E)^{-1}) \circ \varphi_V^E \circ (\varphi_U^E)^{-1}(x, \xi) \\ &= \varphi_U^F \circ (\varphi_V^F)^{-1} \circ (\varphi_V^F \circ f \circ (\varphi_V^E)^{-1})(x, g_{UV}^E(x).\xi) \\ &= \varphi_U^F \circ (\varphi_V^F)^{-1}(x, u \circ g_{UV}^E(x).\xi) \\ &= (x, g_{UV}^F(x) \circ u \circ g_{VU}^E(x).\xi)\end{aligned}$$

Hence, $\widehat{f}(x) = g_{UV}^F(x) \circ u \circ g_{VU}^E(x)$, and we get

$$\varphi_U \circ \varphi_V^{-1}(x, u) = \varphi_U(f) = (x, \widehat{f_U}(x)) = (x, g_{UV}^F(x) \circ u \circ g_{VU}^E(x)).$$

This shows that the maps $\varphi_U \circ \varphi_V^{-1}$ are smooth. Thus, using the maps φ_U, we can define on $Hom(E, F)$ the structure of a smooth vector bundle of rank rs with transition functions

$$Hom(\mathbb{K}^r, \mathbb{K}^s) \ni u \to g_{UV}^F(x) \circ u \circ g_{VU}^E(x) \in Hom(\mathbb{K}^r, \mathbb{K}^s).$$

If $\underline{\mathbb{K}}^r$ denotes the trivial bundle $M \times \mathbb{K}^r$, the dual bundle E^* can be identified with the bundle $Hom(E, \underline{\mathbb{K}}^r)$ in an obvious way.

5. $\Lambda^p E$ with transitions functions $\Lambda^p g_{UV}(x) \in GL(\Lambda^p \mathbb{K}^r)$.

If the rank of a vector bundle E is r, the bundle $\Lambda^r E$ is called the **determinant bundle** of E and is often denoted by $det\ E$. The reason for this name is the following. Fix a basis $e_1, \ldots, e_r$ of $\mathbb{K}^r$. Then $e_1 \wedge \ldots \wedge e_r$ is a basis of $\Lambda^r \mathbb{K}^r$ and, using this basis, we can identify $\Lambda^r \mathbb{K}^r$ with $\mathbb{K}$. Under this identification, if $A \in Hom(\mathbb{K}^r, \mathbb{K}^r)$, the map $\Lambda^r A$ is identified with the map $\mathbb{K} \ni \lambda \to det\ A \cdot \lambda \in \mathbb{K}$. Note that the latter map does not depend on the choice of the basis of $\mathbb{K}^r$. These remarks allow one to say that $\Lambda^r E$ has as transition functions the maps "the determinant of $g_{\alpha,\beta}$", $det\ g_{\alpha,\beta} : U_\alpha \cap U_\beta \to GL(\mathbb{K}) \cong \mathbb{K} \setminus \{0\}$.

6. $\odot^p E$ with transition functions $\odot^p g_{UV}(x) \in GL(\odot^p \mathbb{K}^r)$.

Note that if M is a smooth manifold, the bundle $(\otimes^r TM) \otimes (\otimes^s T^*M)$ is called the bundle of **tensors** of type (r, s) or tensors of contravariant degree r and covariant degree s.

Problem 1.4.1 *Let* $\underline{\mathbb{K}} = M \times \mathbb{K}$ *be the trivial bundle over* M. *Show that the map* $E \otimes \underline{\mathbb{K}} \to E$ *defined by* $v \otimes \lambda \to \lambda v$ *is an isomorphism.*

Problem 1.4.2 *Let* $\pi_E : E \to M$ *and* $\pi_F : F \to M$ *be vector bundles. Show that the set* $\{(a, b) \in E \times F : \pi_E(a) = \pi_F(b)\}$ *admits a natural structure of a vector bundle over* M *isomorphic to the bundle* $E \oplus F$.

Problem 1.4.3 *Prove that* $Hom(E, F) \cong E^* \otimes F$. *More precisely, there is a unique isomorphism* $\varphi : E^* \otimes F \to Hom(E, F)$ *such that if* $\alpha \in E^*$ *and* $f \in F$, *then* $\varphi(\alpha \otimes f)(a) = \alpha(a) f$ *for every* $a \in E$.

Problem 1.4.4 *Let* $u : E \to F$ *be a morphism of vector bundles over a manifold* M. *Denote by* $u^* : F^* \to E^*$ *the map that on each fibre is the dual map* $u^* : F_x^* \to E_x^*$ *of the linear map* $u : E_x \to F_x$. *Prove that the map* u^* *is smooth, hence it is a morphism of vector bundles.*

If E is a real vector bundle, the real vector bundle $E \otimes \underline{\mathbb{C}}$ possesses a natural structure of a complex vector bundle: the multiplication by the imaginary unit i on the fibres is $i(v \otimes z) = v \otimes iz$, $v \in E_x$, $z \in \underline{\mathbb{C}}_x = \mathbb{C}$. This bundle is called the **complexification** of E and is usually denoted by $E^{\mathbb{C}}$.

Problem 1.4.5 *If* E *is a real vector bundle, then* $E \oplus E$ *admits a natural structure of a complex vector bundle defined by* $i(v, w) = (-w, v)$ *for* $v, w \in E_x$. *Show that there is a canonical isomorphism* $E^{\mathbb{C}} \cong E \oplus E$.

Problem 1.4.6 *A complex vector bundle* E *is the complexification of a real vector bundle (up to isomorphism) exactly when there exists a morphism* $\gamma : E \to E$ *of the real vector bundle* E *such that* $\gamma^2 = Id$ *and* $\gamma(ie) = -i(\gamma(e))$ *for every* $e \in E$.

Now, let E be a complex vector bundle. On every fibre E_x, define a new structure of a complex vector space setting $i.v = -iv$, $v \in E_x$, i.e., $z.v = \overline{z}v$ for $z \in \mathbb{C}$. In this way, we obtain a new complex vector bundle often denoted by $\overline{E}$ and called the conjugate of E. The bundles E and $\overline{E}$ are identical as real vector bundles, but they may not be isomorphic as complex vector bundles. However, if E is the complexification of a real vector bundle F, $E = F^{\mathbb{C}}(= F \oplus F)$, the map $v + iw \to v - iw = v + i.w$, $v, w \in F_x$, is an isomorphism of the complex vector bundles E and $\overline{E}$. In this case, the conjugate element $v - iw$ of $a = v + iw$ will be denoted by $\overline{a}$ as usual.

Let $\varphi = (\pi, \Phi)$ be a trivialization of the complex vector bundle E, where π is the projection of E. Then $\overline{\varphi} = (\pi, \overline{\Phi})$ is a trivialization of $\overline{E}$, the bar on the right-hand side meaning complex conjugation in $\mathbb{C}^r$. Therefore, if g_{UV} are transition functions of E, then $\overline{g}_{UV}$ are transition functions of $\overline{E}$.

Let E and F be real vector bundles. Prove that:

Problem 1.4.7 *There exists a unique canonical isomorphism $(E \otimes F)^{\mathbb{C}} \cong E^{\mathbb{C}} \otimes F^{\mathbb{C}}$ such that $v_1 \otimes_{\mathbb{R}} w_1 + i v_2 \otimes_{\mathbb{R}} w_2 \to v_1 \otimes_{\mathbb{C}} w_1 + i v_2 \otimes_{\mathbb{C}} w_2$, where the complexification of a real bundle is understood as in Problem 1.4.5.*

Problem 1.4.8 *The map $(E^*)^{\mathbb{C}} \to (E^{\mathbb{C}})^*$ defined by $\alpha + i\beta \to \varphi$ where $\varphi(v + iw) = [\alpha(v) - \beta(w)] + i[\alpha(w) + \beta(v)]$ is an isomorphism.*

These isomorphisms imply $((E \otimes F)^*)^{\mathbb{C}} \cong (E^{\mathbb{C}} \otimes F^{\mathbb{C}})^*$. In particular, $(E^{\mathbb{C}} \otimes F^{\mathbb{C}})^*$ is the complexification of a real vector bundle.

Problem 1.4.9 *Let $L_2(E_x^{\mathbb{C}}, F_x^{\mathbb{C}}; \mathbb{C})$ be the space of bilinear forms $E_x^{\mathbb{C}} \times F_x^{\mathbb{C}} \to \mathbb{C}$. Consider the standard isomorphism $(E_x^{\mathbb{C}} \otimes F_x^{\mathbb{C}})^* \cong L_2(E_x^{\mathbb{C}}, F_x^{\mathbb{C}}; \mathbb{C})$ and the resulting isomorphism $((E_x \otimes F_x)^*)^{\mathbb{C}} \cong L_2(E_x^{\mathbb{C}}, F_x^{\mathbb{C}}; \mathbb{C})$. Let $\Psi \in ((E_x \otimes F_x)^*)^{\mathbb{C}}$ and let $\psi = \alpha + i\beta$ be the corresponding bilinear form with $\alpha(v, w) = Re\, \psi(v, w)$, $\beta(v, w) = Im\, \psi(v, w)$. Show that the bilinear form corresponding to the conjugate element $\overline{\Psi}$ is $\alpha - i\beta$, and denoting the latter form by $\overline{\psi}$, we have $\overline{\psi}(v, w) = \overline{\psi(\overline{v}, \overline{w})}$ for $u \in E_x^{\mathbb{C}}$, $v \in F_x^{\mathbb{C}}$.*

1.5 Pull-Back of a Vector Bundle

Let $f : M \to N$ be a smooth map of smooth manifolds. Let further $\pi : E \to N$ be a smooth vector bundle over N. The pull-back of E under f is the bundle f^*E over M whose fibre at a point $x \in M$ is the vector space $E_{f(x)}$. Strictly speaking

$$f^*E = \{(x, v) \in M \times E : f(x) = \pi(v)\}.$$

The projection $\widetilde{\pi} : f^*E \to M$ is defined by $(x, v) \to x$. Let $\varphi_U : \pi^{-1}(U) \to U \times \mathbb{K}^r$ be a trivialization of E over an open set U, and let $p_2 : U \times \mathbb{K}^r \to \mathbb{K}^r$ be the projection on the second factor. Then $\widetilde{U} = f^{-1}(U)$ is an open subset of M and, on the set $\widetilde{\pi}^{-1}(\widetilde{U})$, we define the map $\widetilde{\varphi}_{\widetilde{U}}(x, v) = (x, p_2 \circ \varphi_U(v))$. This map is a bijection of $\widetilde{\pi}^{-1}(\widetilde{U})$ onto $\widetilde{U} \times \mathbb{K}^r$; the inverse map is given by $\widetilde{\varphi}_{\widetilde{U}}^{-1}(x, \xi) = (x, \varphi_U^{-1}(f(x), \xi))$. Using the maps $\widetilde{\varphi}_{\widetilde{U}}$, we can define on f^*E the structure of a smooth vector bundle with transition functions $g_{\widetilde{U}\widetilde{V}} = g_{UV} \circ f\ (= f^*g_{UV})$. If $\iota : f^*E \hookrightarrow M \times E$ is the inclusion map, we have $(Id_{\widetilde{U}} \times \varphi_U) \circ \iota \circ \widetilde{\varphi}_{\widetilde{U}}^{-1}(x, \xi) = (x, f(x), \xi)$. It follows that ι is a smooth map of rank $dim\, M + r$, hence f^*E is a submanifold of $M \times E$. The topology of this submanifold coincides with the one induced by $M \times E$ (see Problem 1.5.3 below).

Let $\widetilde{f} : f^*E \to E$ be the map $(x, v) \to v$. The identity $\varphi_U \circ \widetilde{f} \circ \widetilde{\varphi}_{\widetilde{U}}^{-1}(x, \xi) = (f(x), \xi)$ implies that the map $\widetilde{f}$ is smooth. It sends the fibre of f^*E at a point x onto $E_{f(x)}$ and is a linear isomorphism. The existence of a smooth map $\widetilde{f} : f^*E \to E$

whose restriction to any fibre is a linear isomorphism uniquely determines the pull-back bundle up to isomorphism. Indeed, let $\pi' : E' \to M$ be a bundle over M and suppose that there exists a smooth map $f' : E' \to E$ such that the following diagramme is commutative, i.e., f' sends a fibre into a fibre.

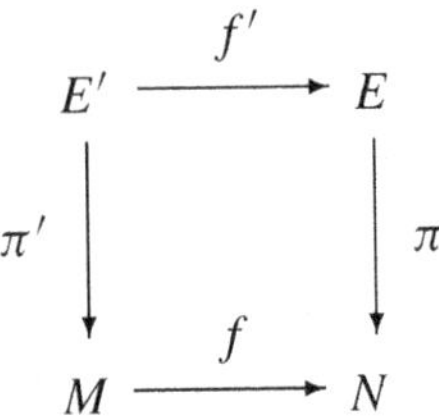

Also, suppose that f' is a linear isomorphism on each fibre. Then the map $h : E' \to f^*E$ defined by $h(w) = (\pi'(w), f'(w))$ is an isomorphism.

Problem 1.5.1 *Prove that if $f : M \to N$ is a constant map, $f(x) \equiv const$, the pull-back bundle $f^\star E$ is isomorphic to the trivial bundle $M \times \mathbb{K}$.*

Problem 1.5.2 *Let $E \to N$ be a smooth vector bundle over a manifold N. Let $h : M' \to M$ and $f : M \to N$ be smooth maps. Show that $(f \circ h)^* E \cong h^*(f^*E)$.*

Problem 1.5.3 *Let $f : M \to N$ be a smooth map and $\pi : E \to N$ a smooth vector bundle over N.*

Let $\psi : M \times E \to N \times N$ be the map $\psi(x, v) = (f(x), \pi(v))$. The diagonal $\widetilde{N} = \{(p, q) \in N \times N : p = q\}$ is an embedded submanifold of $N \times N$, and $f^*E = \psi^{-1}(\widetilde{N})$.

(a) Prove that the map ψ is transversal to the submanifold $\widetilde{N}$, i.e., for every $a = (x, v) \in f^*E$,

$$T_{\psi(a)}(N \times N) = \psi_*(T_a(M \times E)) + T_{\psi(a)}\widetilde{N} \text{ (the sum is not required to be direct)} .$$

It follows that the set f^*E possesses the structure of an embedded submanifold $\widetilde{E}'$ of $M \times E$ of dimension $dim\,(M \times E) - dim\,(N \times N) + dim\,\widetilde{N} = dim\,M + r$ (Warner [24], Chap. 1, Theorem 1.39 or Narasimhan [20], Chap. 2, § 2.16, Proposition 2.16.3).

Let $\jmath : \widetilde{E}' \hookrightarrow M \times E$ be the inclusion map.

(b) Prove that the identity map $Id : f^*E \to \widetilde{E}'$ is a diffeomorphism making use of the identity $\imath = \jmath \circ Id$, where $\imath : f^*E \hookrightarrow M \times E$ is the inclusion map.

This means that the manifolds f^*E and $\widetilde{E}'$ coincide.

Problem 1.5.4 *Let $E \to N$ be a real vector bundle and $f : M \to N$ a smooth map. Prove that the complexification of the pull-back bundle $(f^*E)^{\mathbb{C}}$ is isomorphic to the pull-back bundle $f^*E^{\mathbb{C}}$ (in a canonical way).*

Example 1.5.1 Let N be a submanifold of M and $i : N \hookrightarrow M$ the inclusion map. Let $\pi : E \to M$ be a smooth vector bundle. Then $i^*E = \{(x, v) \in N \times E : v \in E_x\}$ and the map $(x, v) \to v$ is an isomorphism of i^*E onto the restriction $E|N = \pi^{-1}(N)$ of the bundle E to the submanifold N.

Now, recall that two smooth maps $f_0, f_1 : M \to N$ between smooth manifolds are called homotopic, if there exists a smooth map $F : [0, 1] \times M \to N$ such that $F(0, x) = f_0(x)$ and $F(1, x) = f_1(x)$.

The proof of the next result can be seen, for example, in Husemoller [13], Part 1, Chap. 3, Sect. 4, Theorem 4.7 or in Hirsch [11], Chap. 4, Sect. 2, Theorem 2.4.

Proposition 1.5.1 *Let $E \to N$ be a smooth vector bundle over a smooth manifold N. If M is a paracompact manifold and $f_0, f_1 : M \to N$ are smooth homotopic maps, then $f_0^*E \cong f_1^*E$.*

1.6 Sections of Vector Bundles. Frames

Definition 1.6.1 Let $\pi : E \to M$ be a smooth vector bundle and U an open subset of M. A section of E on U is a smooth map $\sigma : U \to E$ such that $\pi \circ \sigma = Id$, i.e., $\sigma(x)$ belongs to the fibre E_x for every $x \in U$.

Sometimes the image $\sigma(U)$ is also called a section of E. The maps $\sigma : U \to E$ with the property $\pi \circ \sigma = Id$ are often called sections of E even if they are not smooth or continuous. It is usually clear from the context whether the sections are smooth, continuous, or arbitrary.

Note that if $f : U \to \mathbb{K}$ is a smooth function and $s : U \to E$ a smooth section, the section $fs : U \ni x \to f(x)s(x) \in E$ is also smooth. Indeed, let $p \in U$ and let $V \subset U$ be a trivializing neighbourhood of p with trivialization $\varphi = (\pi, \Phi) : \pi^{-1}(V) \to V \times \mathbb{K}^r$. Then $\varphi \circ (fs) = (\pi, f(\Phi \circ s))$ is a smooth map. This proves the smoothness of fs. If $\sigma : U \to E$ is also a smooth section, the section $s + \sigma$ is smooth since $\varphi \circ (s + \sigma) = (\pi, \Phi \circ s + \Phi \circ \sigma))$. Therefore the space of smooth sections on U is a module over the ring of smooth functions on U.

Problem 1.6.1 *Prove that the pre-image $s^{-1}(Q)$ of a compact subset Q of E under a section s of E is a compact subset of M.*

Example 1.6.1 The map $x \to 0_x \in E_x$ is called the zero section.

Example 1.6.2 Let M be a smooth manifold. An arbitrary section of the trivial bundle $M \times \mathbb{K}^r$ is of the form $x \to (x, f(x))$ where $f : M \to \mathbb{K}^r$. Thus, the smooth sections of the trivial bundle can be identified with the smooth functions $f : M \to \mathbb{K}^r$.

Example 1.6.3 Let $E \to M$ be a $\mathbb{K}$-vector bundle. The sections of the bundle $\otimes^p E^*$ can be identified with the maps assigning a $\mathbb{K}$-multilinear form $E_x \times \ldots \times E_x$ (p times)$\to \mathbb{K}$ to each point $x \in M$. The sections of $\odot^p E^*$ correspond to symmetric p-forms on the fibres of E, those of $\Lambda^p E^*$ to skew-symmetric forms.

Example 1.6.4 Every smooth section X of the tangent bundle TM of a smooth manifold M is, in fact, a smooth vector field on M. The smoothness condition for X is equivalent to the condition that for every point p of M there exist local coordinates $x_1, \ldots, x_n$ of M in a neighbourhood of p such that if X is represented as $X = \sum_{i=1}^n f_i \dfrac{\partial}{\partial x_i}$, then the functions f_i are smooth. If this property is satisfied for a local coordinate system $x_1, \ldots, x_n$, it is satisfied for any other coordinate system.

Each section of the complexified tangent bundle $T^{\mathbb{C}}M$ is a complex vector field on M; it is of the form $X + iY$ where X and Y are vector fields on M.

Example 1.6.5 Each smooth section ω of the bundle $\Lambda^k T^*M$ is a k-form on M. The smoothness condition for ω is equivalent to the condition that for every point p of M there exist local coordinates $x_1, \ldots, x_n$ of M in a neighbourhood of p such that ω has the representation

$$\omega = \sum_{1 \le i_1 < \ldots < i_k \le n} f_{i_1 \ldots i_k} \mathrm{d}x_{i_1} \wedge \ldots \wedge \mathrm{d}x_{i_k}$$

where the functions $f_{i_1 \ldots i_k}$ are smooth.

Example 1.6.6 Let $E \to M$ be a smooth vector bundle. Every section of the bundle $\Lambda^k T^*M \otimes E$ is a k-form with values in E. If ω is such a form, for every $x \in M$, the value of ω at x is a k-multilinear skew-symmetric map $\omega_x : T_xM \times \ldots \times T_xM \to E_x$.

Let $\Gamma(TM)$ and $\Gamma(E)$ be the spaces of smooth (global) sections of TM and E. Every form with values in E can be considered as a skew-symmetric map

$$\widetilde{\omega} : \Gamma(TM) \times \ldots \times \Gamma(TM) \to \Gamma(E),$$

which is multilinear over the smooth functions on M:

$$\widetilde{\omega}(f_1 X_1, \ldots, f_k X_k) = f_1 \ldots f_k \widetilde{\omega}(X_1, \ldots X_k),$$

for $X_1, \ldots, X_k \in \Gamma(TM)$ and smooth functions $f_1, \ldots, f_k$ on M. If ω is a k-form with values in E, the map corresponding to ω is defined by $\widetilde{\omega}(X_1, \ldots., X_k)(x) = \omega(X_1(x), \ldots, X_k(x))$. Conversely, let $\widetilde{\omega}$ be a map having the properties mentioned above. For any $v_1, \ldots, v_k \in T_xM$, there exist vector fields $X_1, \ldots, X_k$ on M such that $X_1(x) = v_1, \ldots, X_k(x) = v_k$ (see Remark 2.1.1 in Chap. 2, if necessary). It is a key observation that $\widetilde{\omega}(X_1, \ldots, X_k)(x)$ does not depend on the choice of the vector fields $X_1, \ldots, X_k$ with $X_j(x) = v_j$, $j = 1, \ldots, k$. To show this, we need the following claim: If $Z_1, \ldots, Z_k$ are vector fields on M and $Z_i(x) = 0$ for some index i, then $\widetilde{\omega}(Z_1, \ldots, Z_k)(x) = 0$. In order to prove the claim, take a coordinate system

$(V, x_1, \dots x_n)$ of M at the point x. We have $Z_i = \sum_{l=1}^n h_l \frac{\partial}{\partial x_l}$ on V where h_l are smooth functions with $h_l(x) = 0$. Next, we extend the vector fields $\frac{\partial}{\partial x_l}$ on V to global vector fields on M. As it is remarked in the proof of Chap. 2, Lemma 2.1.1, there is a smooth function $\widetilde{\varphi}$ on M such that $\widetilde{\varphi} = 0$ outside a compact subset K of V and $\widetilde{\varphi} = 1$ in a neighbourhood of x. For $l = 1, \dots, n$, the vector field E_l defined by $E_l = \widetilde{\varphi}\frac{\partial}{\partial x_l}$ on V and $E_l = 0$ on $M \setminus V$ is smooth on the open sets V and $M \setminus K$ covering M, hence E_l is smooth on M. The functions $f_l = \widetilde{\varphi} h_l$ on V, $f_l = 0$ on $M \setminus V$ are smooth on M and $f_l(x) = 0$, Clearly, $Z_i = \sum_{l=1}^n f_l E_l + (1 - \widetilde{\varphi}^2) Z_i$. Then $\widetilde{\omega}(Z_1, \dots, Z_k)(x) = \sum_{l=1}^n f_l(x)\widetilde{\omega}(Z_1, \dots, E_l, \dots, Z_k)(x) + (1 - \widetilde{\varphi}(x)^2)\widetilde{\omega}(Z_1, \dots, Z_i, \dots, Z_k)(x) = 0$. Now, if $Y_1, \dots, Y_k$ are vector fields on M with $Y_j(x) = v_j$, then $\widetilde{\omega}(X_1 - Y_1, X_2, \dots, X_k)(x) = 0$ by the claim above. Hence, $\widetilde{\omega}(X_1, X_2, \dots, X_k)(x) = \widetilde{\omega}(Y_1, X_2, \dots, X_k)(x)$. Similarly, $\widetilde{\omega}(Y_1, X_2, X_3, \dots, X_k)(x) = \widetilde{\omega}(Y_1, Y_2, X_3, \dots, X_k)(x)$, and we see by induction that $\widetilde{\omega}(X_1, \dots, X_k)(x) = \widetilde{\omega}(Y_1, \dots, Y_k)(x)$. Therefore the identity $\omega(v_1, \dots, v_k) = \widetilde{\omega}(X_1, \dots, X_k)(x)$ correctly defines a k-form with values in E

If E is the trivial bundle, $\Lambda^k T^*M \otimes E \cong \Lambda^k T^*M$, and the forms with values in E are just the usual differential forms on M.

Example 1.6.7 Let $\pi_E : E \to M$ and $\pi_F : F \to M$ be smooth vector bundles.

* Every section of the bundle $E \oplus F$ is of the form $s + \sigma$, where s is a section of E and σ is a section of F.

* Every section σ of the bundle $Hom(E, F)$ can be identified with a morphism $\widetilde{\sigma} : E \to F$. If σ is a section of $Hom(E, F)$, the morphism $\widetilde{\sigma}$ is defined by $\widetilde{\sigma}(e) = \sigma(\pi_E(e))(e)$ for $e \in E$. Conversely, if $\widetilde{\sigma} : E \to F$ is a morphism, we set $\sigma(x) = \widetilde{\sigma}|E_x : E_x \to F_x$ for $x \in M$. Let U be an open subset of M on which the bundles E and F are trivial with trivializations φ_U^E and φ_U^F. Then

$$\varphi_U^F \circ \widetilde{\sigma} \circ (\varphi_U^E)^{-1}(x, \xi) = (x, \widehat{\sigma}(x).\xi), \ x \in U, \xi \in \mathbb{K}^s,$$

where the map $\widehat{\sigma} : U \to Hom(\mathbb{K}^r, \mathbb{K}^s)$ is smooth (by Lemma 1.2.1). Let φ_U be the trivialization of $Hom(E, F)$ defined by means of φ_U^E and φ_U^F (Sect. 1.4, item **4.**). Then $\varphi_U \circ \sigma = \widehat{\sigma}$. It follows that the map $\sigma : M \to Hom(E, F)$ is smooth. Thus, σ is a smooth section of $Hom(E, F)$.

Example 1.6.8 Let $f : M \to N$ be a smooth map between smooth manifolds and $E \to N$ a smooth vector bundle over N. Every smooth section $\widetilde{s}$ of the pull-back bundle f^*E over an open set $U \subset M$ has the form $\widetilde{s}(x) = (x, s(x))$, where $s : U \to E$ is a smooth map such that $s(x) \in E_{f(x)}$. If $\widetilde{f} : f^*E \to E$ is the standard map which sends a fibre into a fibre and is a vector space isomorphism on the fibres, then $s = \widetilde{f} \circ \widetilde{s}$ (hence $\widetilde{s}(x) = (\widetilde{f}|E_{f(x)})^{-1}(s(x))$). A map s with this property is said to be a section of E along the map f. Thus, the sections of f^*E can be identified with the sections of E along f.

Example 1.6.9 Let $E \to M$ be a smooth complex vector bundle. For every $x \in M$, define a map $J_x : E_x \to E_x$ setting $J_x(v) = iv$. Consider E as a real vector bundle; for the moment, denote this real vector bundle by E^{re}. Then the map $x \to J_x$ is a smooth section of the bundle $Hom(E^{re}, E^{re})$ (equivalently, a morphism $E^{re} \to E^{re}$) with the property $J_x^2 = -Id$. This section is called the complex structure of the complex vector bundle E.

Conversely, let $\widetilde{E}$ be a smooth real vector bundle possessing a section J of the bundle $Hom(\widetilde{E}, \widetilde{E})$ such that $J^2 = -Id$. Then we can define a structure of a complex vector space on every fibre of $\widetilde{E}$ setting $iv = J(v)$, i.e., $(a + ib)v = av + bJ(v)$, $a, b \in \mathbb{R}$. In this way, we obtain a smooth complex vector bundle E for which $E^{re} = \widetilde{E}$.

Thus, the notion of a smooth complex vector bundle is equivalent to the notion of a smooth real vector bundle endowed with a morphism J such that $J^2 = -Id$ (of course, the latter identity implies that J is an isomorphism).

Now, let $E = T^h M$ be the holomorphic tangent bundle of a complex manifold M. As we have noted, $T^h M$ as a real vector bundle is isomorphic to the tangent bundle TM of M considered as a smooth real manifold. Therefore the complex structure of the bundle $T^h M$ determines a morphism J of the (real) bundle TM with the property $J^2 = -Id$. Let $z_k = x_k + iy_k$, $k = 1, \ldots, n$, be complex coordinates on M. Under the isomorphism of TM with $T^h M$, we have $\frac{\partial}{\partial x_k} \to \frac{\partial}{\partial z_k}$ and $\frac{\partial}{\partial y_k} \to i\frac{\partial}{\partial z_k}$. Hence

$$J\frac{\partial}{\partial x_k} = \frac{\partial}{\partial y_k}, \quad J\frac{\partial}{\partial y_k} = -\frac{\partial}{\partial x_k}. \tag{1.8}$$

A smooth manifold whose tangent bundle admits a morphism J such that $J^2 = -Id$ is called **almost complex**; the morphism J is called an **almost complex structure**. Note that the smoothness condition of J is equivalent to the condition that for every point $p \in M$ there exist local coordinates $x_1, \ldots, x_n$ in a neighbourhood of p such that if $J\frac{\partial}{\partial x_k} = \sum_{l=1}^{n} J_{kl}\frac{\partial}{\partial x_l}$, then the functions J_{kl} are smooth.

Recall that a map $f : M \to N$ between complex manifolds is holomorphic if for every point $p \in M$ there exist coordinate systems (U, z), $z = (z_1, \ldots, z_m)$, of M at the point p and (V, w), $w = (w_1, \ldots, w_n)$, of N at $f(p)$ such that $f(U) \subset V$, the functions $w_l \circ f$, $l = 1, \ldots, n$, are smooth on U, and if $z_k = x_k + iy_k$, $1 \leq k \leq m$ and $w_l = u_l + iv_l$, $1 \leq l \leq n$, then the Cauchy-Riemann equations hold: $\frac{\partial(u_l \circ f)}{\partial x_k} = \frac{\partial(v_l \circ f)}{\partial y_k}$, $\frac{\partial(u_l \circ f)}{\partial y_k} = -\frac{\partial(v_l \circ f)}{\partial x_k}$. If f is holomorphic these equations hold for every two coordinate systems (U, z) and (V, w) with $f(U) \subset V$. Let J be the complex structure of M, and I the complex structure of N. It follows from (1.8) and the similar identity for I that the Cauchy-Riemann system of equations holds exactly when $f_*(JX) = If_*(X)$ for every tangent vector $X \in TM$. Thus, a map $f : M \to N$ is holomorphic exactly when it is smooth and satisfies the identity $f_* \circ J = I \circ f_*$. In this form, the condition for holomorphicity of f makes sense for almost complex manifolds. Often holomorphic maps between almost complex manifolds are called pseudo holomorphic.

A natural question is when given a smooth manifold M endowed with an almost complex structure J one can introduce a complex manifold structure on the set M such that $J(X) = iX$, i.e., if $z_k = x_k + iy_k$, $k = 1, \ldots, n$, are complex coordinates, then $x_1, y_1, \ldots x_n, y_n$ are coordinates of the smooth manifold M and $J\frac{\partial}{\partial x_k} = \frac{\partial}{\partial y_k}$, $J\frac{\partial}{\partial y_k} = -\frac{\partial}{\partial x_k}$. The answer to this question is given by a theorem of Newlander and Nirenberg. To state this theorem, we need to introduce the so-called Nijenhuis tensor of J. Let $p \in M$ and let X, Y be (smooth) vector fields on M in a neighbourhood of the point p. Set

$$N^J(X, Y) = -[X, Y] + [JX, JY] - J([JX, Y] + [X, JY]).$$

It is easy to check that $N^J(X, Y)$ is bilinear over the ring of smooth functions. Hence, the values of $N^J(X, Y)$ at the point p depends only on the values of X and Y at p. This allows one to define a bilinear map $N_p^J : T_pM \times T_pM \to T_pM$ in the following way. Let $v, w \in T_pM$, and take vector fields X, Y in a neighbourhood of the point p such that $X_p = v$, $Y_p = w$. Vector fields with this property always exist. For example, if $x_1, \ldots, x_n$ are local coordinates of M in a neighbourhood of p, then $v = \sum_{i=1}^n v_i \frac{\partial}{\partial x_i}(p)$, $v_i \in \mathbb{R}$, and we can set $X = \sum_{i=1}^n v_i \frac{\partial}{\partial x_i}$; similarly for Y. Since $N^J(X, Y)(p)$ does not depend on the choice of X and Y, the identity $N^J(v, w) = N^J(X, Y)(p)$ correctly defines the map N_p^J. By the very definition, N_p^J is a skew-symmetric bilinear map smoothly depending on the point p, i.e., N^J is a section of the bundle $\Lambda^2 T^*M \otimes TM$.

Proposition 1.6.1 *An almost complex structure J on a smooth manifold M originates from a complex-analytic atlas (i.e., it is a complex structure) if and only if $N^J = 0$. Any two such atlases determine the same structure of a complex manifold on M.*

Problem 1.6.2 *Prove that $N^J = 0$ if J is the complex structure of a complex manifold.*

An almost complex structure with vanishing Nijenhuis tensor is called **integrable.** In dimension 2, every almost complex structure J is integrable; this follows from the fact that if X is a non-zero tangent vector, then X, JX is a basis of the tangent space.

Comments on the Newlander-Nirenberg theorem and other facts about complex and almost complex manifolds can be found in Kobayashi and Nomizu [14] vol. II, Chap. IX and Appendix 8. A version of this theorem in terms of differential forms can be seen in Narasimhan [20], Chap. 2, § 2.12.

Definition 1.6.2 Let $\pi : E \to M$ be a smooth vector bundle and U an open subset of M. A set of smooth sections $s_1, \ldots, s_r$ of E over U is called a (local) frame if for every point $x \in U$ the vectors $s_1(x), \dots, s_r(x)$ constitute a basis of the vector space E_x.

The notions of local frames and local trivializations are equivalent. Indeed, if $\varphi_U : \pi^{-1}(U) \to U \times \mathbb{K}^r$ is a local trivialization of E, take a basis $a_1, \dots, a_r$ of $\mathbb{K}^r$ and set $s_i(x) = \varphi_U^{-1}(x, a_i),\ x \in U,\ i = 1, \dots, r$. Then $s_1, \dots, s_r$ is a frame of E over U. Conversely, let $s_1, \dots, s_r$ be a frame on U. Then every $e \in \pi^{-1}(U)$ can be uniquely written as

$$e = \sum_{i=1}^{r} \lambda_i(e) s_i(\pi(e)),\ \lambda_i(e) \in \mathbb{K}. \tag{1.9}$$

Take a basis $a_1, \dots, a_r$ of $\mathbb{K}^r$ and set

$$\varphi_U(e) = (\pi(e), \sum_{i=1}^{r} \lambda_i(e) a_i) \in U \times \mathbb{K}^r.$$

Then φ_U is a bijective map of $\pi^{-1}(U)$ onto $U \times \mathbb{K}^r$ sending every fibre $E_x,\ x \in U$, isomorphically onto the vector space $\{x\} \times \mathbb{K}^r$. Clearly, $s_i(x) = \varphi_U^{-1}(x, a_i),\ i = 1, \dots, r$. It remains to prove that φ_U is a diffeomorphism. Let $e_0 \in \pi^{-1}(U),\ x_0 = \pi(e_0)$, and let $V \subset U$ be a neighbourhood of the point x_0 for which there exists a local trivialization $\psi = (\pi, \Psi) : \pi^{-1}(V) \to V \times \mathbb{K}^r$. The functions $f_i = \Psi \circ s_i : V \to \mathbb{K}^r$ are smooth, $i = 1, \dots, r$, and $f_i(x) = \Psi(s_i(x))$ is a basis of $\mathbb{K}^r$ for every $x \in V$. For $x \in V$ and $\xi = \sum_{i=1}^r \xi_i a_i \in \mathbb{K}^r$, we have $\varphi_U^{-1}(x, \xi) = \sum_{i=1}^r \xi_i s_i(x)$ and $\psi \circ \varphi_U^{-1}(x, \xi) = (x, \sum_{i=1}^r \xi_i f_i(x))$. It follows that the map $\psi \circ \varphi_U^{-1} | (V \times \mathbb{K}^r)$ is smooth. Therefore the map φ_U^{-1} is smooth. To show that the map φ_U is smooth, it suffices to show that the functions $\lambda_i,\ i = 1, \dots, r$, defined by (1.9) are smooth. Let $\Psi = \sum_{j=1}^r g_j a_j$. The functions g_j on $\pi^{-1}(V)$ in this representation of Ψ are smooth. On the other hand, by (1.9), for every $e \in \pi^{-1}(V)$, we have $\Psi(e) = \sum_{i=1}^r \lambda_i \Psi(s_i(\pi(e)) = \sum_{i=1}^r \lambda_i f_i(\pi(e))$. Hence

$$\sum_{i=1}^{r} \lambda_i (f_i \circ \pi) = \sum_{j=1}^{r} g_j a_j. \tag{1.10}$$

Let $f_i = \sum_{j=1}^r \mu_{ij} a_j$. The functions $\mu_{ij} : V \to \mathbb{K}$ are smooth (since f_i's are smooth) and $\det[\mu_{ij}] \neq 0$ at every point x of V (since $\{f_i(x)\}$ and $\{a_i\}$ are bases of $\mathbb{K}^r$). If $[\mu^{ij}]$ is the inverse matrix of $[\mu_{ij}]$, the functions μ^{ij} are smooth by the standard rule for finding an inverse matrix. We have $a_j = \sum_{i=1}^r \mu^{ji} f_i$, thus by (1.10), $\lambda_i = \sum_{j=1}^r g_j (\mu^{ji} \circ \pi)$. Therefore the functions λ_i are smooth.

It follows from the equivalence of the notions of frames and trivializations that a bundle is isomorphic to the trivial one if and only if it admits a globally defined frame.

Using this fact, it is easy to see that if L is a line bundle and L^* is its dual bundle, the bundle $L \otimes L^*$ is trivial. Indeed, let s be a local frame of L (i.e., a non-vanishing section) and let s^* be its dual frame. Then $s \otimes s^*$ is a frame of the bundle $L \otimes L^*$. If σ is another frame of L whose domain intersects the domain of s, then $\sigma = fs$, where f is a non-vanishing smooth function. The dual frame σ^* is $\sigma^* = \frac{1}{f}s^*$, hence $\sigma \otimes \sigma^* = s \otimes s^*$. This means that the local frames $s \otimes s^*$ glue together to define a global frame of $L \otimes L^*$. Hence, this bundle is trivial. Thus, the set of isomorphic classes of line bundles with the operation "tensor product" has the structure of a group. It is called the Picard group.

A smooth manifold admitting a global frame of vector fields, i.e., with trivial tangent bundle, is called **parallelizable**.

Problem 1.6.3 *Prove that the spheres S^1, S^3, S^7 are parallelizable manifolds.*

Remark 1.6.1 These are the only parallelizable spheres as proved by Bott and Milnor [1].

Another consequence of the equivalence of frames and local trivializations is that a map $f : E \to F$ of vector bundles over a manifold M sending a fibre into a fibre is smooth exactly when in a neighbourhood of every point M there exist frames $s_1, \dots, s_r$ of E and $\sigma_1, \dots, \sigma_m$ of F such that the coefficients λ_{ij} in the sum $f(s_i) = \sum_{j=1}^{m} \lambda_{ij}\sigma_j$, $i = 1, \dots, r$, are smooth functions (if the map f is smooth this holds for every frame of E and every frame of F). The following condition for smoothness of a section φ of the bundle $(\otimes^k E^*) \otimes F$, where E and F are smooth vector bundles over a manifold M, also follows from the equivalence of frames and local trivializations. Recall first that for every $x \in M$ the space $(\otimes^k E_x^*) \otimes F_x$ is canonically isomorphic to the space of multilinear maps $E_x \times \dots \times E_x$ (k times) $\to F_x$. Consider the values $\varphi_x = \varphi(x)$ of φ at the points of M as such maps. Then the smoothness condition for φ is equivalent to the condition that for every $p \in M$ there exists a frame $s_1, \dots, s_r$ of E in a neighbourhood of the point p such that the section $x \to \varphi_x(s_{i_1}(x), \dots, s_{i_k}(x))$ of F is smooth for every $i_1, \dots, i_k \in \{1, 2, \dots, r\}$. The latter condition is satisfied exactly when for every set of local sections $\sigma_1, \dots, \sigma_k$ of E the map $x \to \varphi_x(\sigma_1, \dots, \sigma_k)$ is a smooth section of F.

Let $e_U = (e_1, \dots, e_r)$ and $e_V = (e'_1, \dots, e'_r)$ be frames of E over open subsets U and V, $U \cap V \neq \emptyset$. Let $e'_i = \sum_{j=1}^{r} g_{ij} e_j$, $i = 1, \dots, r$, or in matrix form $e_V = e_U.[g_{ij}]$ (the matrix acts by multiplication on the right of a row-vector). Take bases $a = (a_1, \dots, a_r)$ and $a' = (a'_1, \dots, a'_r)$ of $\mathbb{K}^r$. These bases and the frames e_U, e_V define trivializations φ_U and φ_V of the bundle E. Denote by g_{UV} the transition function determined by these trivializations : $\varphi_U \circ \varphi_V^{-1}(x, \xi) = (x, g_{UV}(x).\xi)$. Then

$$\varphi_U \circ \varphi_V^{-1}(x, a'_i) = \varphi_U(e'_i) = (x, \sum_{j=1}^{r} g_{ij}(x) a_j). \tag{1.11}$$

Hence, the matrix of $g_{UV}(x) \in GL(\mathbb{K}^r)$ with respect to the bases a' and a of $\mathbb{K}^r$ is $[g_{ij}(x)]$, the transition matrix from the basis $e_V(x)$ to the basis $e_U(x)$ of the vector space E_x. Conversely, let $e_i(x) = \varphi_U^{-1}(x, a_i)$ and $e'_i(x) = \varphi_V^{-1}(x, a_i)$, $i = 1, \ldots, r$, be frames determined by trivializations φ_U^{-1} and φ_V^{-1} of E, and bases a, a' of $\mathbb{K}^r$. Then it follows from (1.11) that if $g_{ij}(x)$ are the entries of the matrix of $g_{UV}(x)$ with respect to the bases a' and a, then $e'_i = \sum_{j=1}^r g_{ij} e_j$, i.e., the transition matrix from $e_V(x)$ to $e_U(x)$ is the matrix of $g_{UV}(x)$.

Next, we discuss local representations of sections and morphisms. If σ is a section of E over $U \cap V$, then $\sigma = \sum_{i=1}^r f_i e_i$ and $\sigma = \sum_{i=1}^r f'_i e'_i$, where f_i and f'_i are smooth functions on $U \cap V$. Thus, $\varphi_U \circ \sigma(x) = (x, \sum_{i=1}^r f_i(x) a_i)$ and $\varphi_V \circ \sigma(x) = (x, \sum_{i=1}^r f'_i(x) a'_i)$. Set $\sigma_U = \sum_{i=1}^r f_i a_i$ and $\sigma_V = \sum_{i=1}^r f'_i a'_i$. Since $f_j = \sum_{i=1}^r f'_i g_{ij}$, where $[g_{ij}(x)]$ is the matrix of the linear map $g_{UV}(x)$ with respect to the bases a and a', we have

$$\sigma_U(x) = g_{UV}(x)(\sigma_V(x)), \quad x \in U \cap V. \tag{1.12}$$

Conversely, let $\sigma_U : U \to \mathbb{K}^r$ and $\sigma_V : V \to \mathbb{K}^r$ be smooth maps satisfying identity (1.12). Then $U \ni x \to \varphi_U^{-1}(x, \sigma_U(x))$ and $V \ni y \to \varphi_V^{-1}(y, \sigma_V(y))$ are two smooth sections of E on U and V, respectively. They coincide on $U \cap V$ according to (1.12). Therefore these sections define a section s of E on $U \cup V$ such that $s_U = \sigma_U$ and $s_V = \sigma_V$.

Let $E_1 \to M$ and $E_2 \to M$ be vector bundles of rank r and s, and let $u : E_1 \to E_2$ be a morphism. Let $\{U_\alpha, \varphi_\alpha\}$ and $\{V_i, \psi_i\}$ be atlases of E_1 and E_2 with transition functions $f_{\alpha\beta}$ and g_{ij}. Also, let $u_{i\alpha} : U_\alpha \cap V_i \to Hom(\mathbb{K}^r, \mathbb{K}^s)$ be the corresponding maps representing the morphism u: $\psi_i \circ u \circ \varphi_\alpha^{-1}(x, \xi) = (x, u_{i\alpha}(x).\xi)$. Suppose the triviaizations φ_α and ψ_i are defined by means of frames $e^\alpha = (e_1^\alpha, \ldots, e_r^\alpha)$ and $f^i = (f_1^i, \ldots, f_s^i)$, and bases $a = (a_1, \ldots, a_r)$ of $\mathbb{K}^r$ and $b = (b_1, \ldots, b_s)$ of $\mathbb{K}^s$. Set $u \circ e_k^\alpha = \sum_{l=1}^s u_{kl}^{i\alpha} f_l^i$ on $U_\alpha \cap V_i$, $k = 1, \ldots, r$. Then, for $x \in U_\alpha \cap V_i$ and $\xi = \sum_{k=1}^r \xi_k a_k$, we have $\psi_i \circ u \circ \varphi_\alpha^{-1}(x, \xi) = (x, \sum_{l=1}^s (\sum_{k=1}^r \xi_k u_{kl}^{i\alpha}(x)) b_l)$. Hence, the local representation $\{u_{i\alpha}(x)\}$ of the morphism u at every point $x \in U_\alpha \cap V_i$ is determined by the matrix $[u_{kl}^{i\alpha}(x)]$ of the linear map $u|(E_1)_x : (E_1)_x \to (E_2)_x$ with respect to the bases $e^\alpha(x)$ of $(E_1)_x$ and $f^i(x)$ of $(E_2)_x$. Recall that by identity (1.4) in Sect. 1.3

$$u_{j\beta}(x) = g_{ji}(x) \circ u_{i\alpha}(x) \circ f_{\alpha\beta}(x), \quad x \in U_\alpha \cap U_\beta \cap V_i \cap V_j.$$

The considerations above show that this formula is nothing else, but the well-known relation between the matrices of a linear map with respect to different bases.

Let s be a section of E_1 represented by the functions $s_\alpha : U_\alpha \to \mathbb{K}^r$, $\varphi_\alpha \circ s(x) = (x, s_\alpha(x))$, and let the section $u \circ s$ be represented by $\sigma_i(x) : V_i \to \mathbb{K}^s$, $\psi_i(x) \circ (u \circ s)(x) = (x, \sigma_i(x))$. Then, for $x \in U_\alpha \cap V_i$,

$$(x, u_{i\alpha}(x).s_\alpha(x)) = \psi_i \circ u \circ \varphi_\alpha^{-1}(x, s_\alpha(x)) = \psi_i \circ (u \circ s)(x) = (x, \sigma_i(x)).$$

Hence, $\sigma_i(x) = u_{i\alpha}(x).s_\alpha(x)$ for $x \in U_\alpha \cap V_i$.

Proposition 1.6.2 *Let $s_1, \ldots, s_k$ be smooth sections of E in a neghbourhood of a point $p \in M$. Let $e_{k+1}, \ldots, e_r$ be elements of the fibre E_p such that*

$s_1(p), \dots, s_k(p), e_{k+1}, \dots, e_r$ is a basis of E_p. Then the sections $s_1, \dots, s_k$ can be supplemented to a frame $s_1, \dots, s_k, s_{k+1}, \dots, s_r$ of E in a neighbourhood of p such that $s_{k+1}(p) = e_{k+1}, \dots, s_r(p) = e_r$.

Proof Note first that for every $e \in E_p$, there exists a section s of E in a neighbourhood of p for which $s(p) = e$. Indeed, let $\sigma_1, \dots, \sigma_r$ be a frame of E in a neighbourhood of p. Then $e = \sum_{i=1}^r \xi_i \sigma_i(p), \xi_i \in \mathbb{K}$, and we can set $s = \sum_{i=1}^r \xi_i \sigma_i$.

Now, let $s_{k+1}, \dots, s_r$ be sections of E such that $s_{k+1}(p) = e_{k+1}, \dots, s_r(p) = e_r$. For every $i = 1, \dots, r$, we have $s_i = \sum_{j=1}^r \lambda_{ij} \sigma_j$, where λ_{ij} are smooth functions with values in $\mathbb{K}$. Since $\det[\lambda_{ij}]_{i,\ j=1}^r \neq 0$ at the point p, this determinant is different from zero in a neighbourhood of p by continuity. Then, for every point q in this neighbourhood, the vectors $s_1(q), \dots, s_r(q)$ form a basis of the space E_q, hence $s_1, \dots, s_r$ is a frame of E. □

Corollary 1.6.1 *Let $p \in M$, and let the vectors $e_1, \dots, e_k \in E_p$ be linearly independent. Then, in a neighbourhood of the point p, there exists a frame $s_1, \dots, s_r$ of E such that $s_1(p) = e_1, \dots, s_k(p) = e_k$.*

1.7 Subbundles. Quotient Bundles

Definition 1.7.1 A bundle $\pi_F : F \to M$ is said to be a subbundle of the bundle $\pi_E : E \to M$ if:

(1) F is a subset of E.

(2) The inclusion map $i : F \hookrightarrow E$ is a morphism of bundles.

The second condition means that $i : F \hookrightarrow E$ is a smooth map and for very $x \in M$ the fibre F_x is a vector subspace of E_x (obviously $F_x = \pi_E^{-1}(x) \cap F$).

Example 1.7.1 Let N be a submanifold of a manifold M and $j : N \hookrightarrow M$ the inclusion map. The map $i = j_* : TN \to TM$ is an injective morphism of bundles. Hence $j_*(TN)$ is a subbundle of the bundle $TM|N$. Further on, the bundle TN will be identified with this subbundle.

Let F be a subbundle of E and $s_1, \dots, s_m$ a frame of F in a neighbourhood of a point $p \in M$. Then $\sigma_1 = i \circ s_1, \dots, s_m = i \circ s_m$ are smooth sections of E. The vectors $\sigma_1(p), \dots, \sigma_m(p)$ are linearly independent, thus the sections $\sigma_1, \dots, \sigma_m$ can be completed to a frame $\sigma_1, \dots, \sigma_r$ of E in a neighbourhood of the point p. Take a basis $a_1, \dots, a_r$ of $\mathbb{K}^r$, and consider the space $\mathbb{K}^m$ as the subspace of $\mathbb{K}^r$ yielded by the vectors $a_1, \dots, a_m$. Let φ_U be the trivialization of E determined by the basis $a_1, \dots, a_r$ and the frame $\sigma_1, \dots, \sigma_r$. Also, let ψ_U be the trivialization of F

determined by $a_1, \ldots, a_m$ and the frame $s_1, \ldots, s_m$. Then $\pi_F^{-1}(U) = \pi_E^{-1}(U) \cap F$ and $\psi_U = \varphi_U|\pi_F^{-1}(U)$. This observation proves the following.

Proposition 1.7.1 *If F is a subbundle of E, then for every point $p \in M$ there exist a neighbourhood U of p and a trivialization $\varphi_U : \pi^{-1}(U) \to U \times \mathbb{K}^r$ of E such that $\psi_U = \varphi_U|(\pi_E^{-1}(U) \cap F) : \pi_F^{-1}(U) \to U \times \mathbb{K}^m$ is a trivialization of F over U* ($\mathbb{K}^m$ is considered as a subspace of $\mathbb{K}^r$).

The trivialization ψ_U of F will be called the restriction to F of the trivialization φ_U of E. If ψ_V is a trivialization of F which is the restriction of a trivialization φ_V of E, then $\psi_U \circ \psi_V^{-1}(x, \xi) = \varphi_U \circ \varphi_V^{-1}(x, \xi) = (x, g_{UV}(x).\xi)$ for $x \in U \cap V$ and $\xi \in \mathbb{K}^m$. Hence the corresponding transition function of F is $g_{UV}^F(x) = g_{UV}(x)|\mathbb{K}^m$. In particular, $g_{UV}(x)$ sends $\mathbb{K}^m$ into $\mathbb{K}^m$.

It follows from Proposition 1.7.1 that if F and F' are subbundles of a bundle E of the same rank m and if $F' = F$ as sets, then the map $Id : F' \to F$ is a diffeomorphism. Note also that the vector spaces F_x and F'_x, $x \in M$, coincide since they coincide as sets and both of them are subspaces of E_x of the same dimension m. Thus, the map $Id : F' \to F$ is an isomorphism on the fibres. Therefore $Id : F' \to F$ is an isomorphism of vector bundles. This proves the following

Corollary 1.7.1 *If a subset F of a vector bundle E admits the structure of a subbunle of a given rank m, this structure is unique.*

Proposition 1.7.2 *If F is a subbundle of E, then:*

(1) F is a submanifold of E.

(2) The topology of F coincides with that induced by E.

Proof Under the notation of Proposition 1.7.1,

$$\varphi_U \circ i \circ \psi_U^{-1}(x, \xi_1, \ldots, \xi_m) = (x, \xi_1, \ldots, \xi_m, 0, \ldots, 0)$$

for $(x, \xi_1, \ldots, \xi_m) \in U \times \mathbb{K}^m$. Hence, the differential i_* of i is of rank $\dim M + m = \dim F$, thus F is a submanifold of E.

The homeomorphism φ_U of $\pi_E^{-1}(U)$ onto $U \times \mathbb{K}^r$ maps $\pi_E^{-1}(U) \cap F = \pi_F^{-1}(U)$ onto the topological subspace $U \times \mathbb{K}^m$ of $U \times \mathbb{K}^r$. Therefore, if we consider $\pi_F^{-1}(U)$ with the topology induced by $\pi_E^{-1}(U)$, i.e., as a topological subspace of $\pi_E^{-1}(U)$, then $\varphi_U|\pi_F^{-1}(U)$ is a homeomorphism of the subspace $\pi_F^{-1}(U)$ onto the subspace $U \times \mathbb{K}^m$. The map $\varphi_U|\pi_F^{-1}(U) = \psi_U$ is a trivialization of F, in particular $\varphi_U|\pi_F^{-1}(U)$ is a homeomorphism of the open subset $\pi_F^{-1}(U)$ of F onto $U \times \mathbb{K}^m$. It follows that the topology of $\pi_F^{-1}(U)$ coincides with the topology induced by $\pi_E^{-1}(U)$. Since the sets $\pi_F^{-1}(U)$ form an open covering of F, it easily follows that the topology of F coincides with that induced by E. □

Proposition 1.7.3 *Let E be a vector bundle and F a subset of E. Suppose that:*

(1) For every point $x \in M$, the set $F_x = F \cap E_x$ possesses the structure of a vector subspace of E_x.
(2) $m = dim\, F_x$ does not depend on the point x.

(3) In a neighbourhood of every point $p \in M$, there exist sections $s_1, \ldots, s_m$ of E such that, for every point x of the neighbourhood, the vectors $s_1(x), \ldots, s_m(x)$ form a basis of F_x.

Then F admits the structure of a subbundle of E such that $s_1, \ldots, s_m$ is a frame of the bundle F.

Proof Complete the section $s_1, \ldots, s_m$ to a frame $s_1, \ldots, s_r$ of E in a neighbourhood U of a point $p \in M$ (Proposition 1.6.2). Take a basis of $\mathbb{K}^r$. This basis and the frame $s_1, \ldots, s_r$ determine a trivialization $\varphi_U = \pi_E^{-1}(U) \to U \times \mathbb{K}^r$ of E which maps $\pi_E^{-1}(U) \cap F$ bijectively onto $U \times \mathbb{K}^m$. Set $\psi_U = \varphi_U \big| \pi_E^{-1}(U) \cap F$. Then, for every two neighbourhoods U and V with $U \cap V \neq \emptyset$,

$$\psi_U \circ \psi_V^{-1}(x, \xi_1, \ldots, \xi_m) = \varphi_U \circ \varphi_V^{-1}(x, \xi_1, \ldots, \xi_m, 0, \ldots, 0).$$

hence the maps $\psi_U \circ \psi_V^{-1}$ are smooth. By Proposition 1.1.1, F admits the structure of a smooth vector bundle with projection map $\pi_F = \pi_E \big| F$ and trivializations $\{\psi_U\}$. It is clear that the bundle F is a subbundle of the bundle E. □

Let $E \to M$ and $F \to M$ be smooth vector bundles and $\varphi : E \to F$ a morphism. Set

$$Ker\, \varphi = \{e \in E : \varphi(e) = 0\}, \qquad Im\, \varphi = \varphi(E).$$

Consider $Ker\, \varphi$ and $Im\, \varphi$ with the restrictions of the projection maps of E and F, respectively. These maps fibre the sets $Ker\, \varphi$ and $Im\, \varphi$ over M; the fibres are the vector spaces $Ker\, \varphi | E_x$ and $\varphi(E_x)$, $x \in M$. In the general case, however, $Ker\, \varphi$ and $Im\, \varphi$ are not bundles since they may not satisfy the local triviality condition. For example, let $E = F = (-1, 1) \times \mathbb{R}$ and $\varphi(t, x) = (t, tx)$. Then $Ker\, \varphi | E_0 = \mathbb{R}$ and $\varphi(E_0) = 0$, while $Ker\, \varphi | E_t = 0$ and $\varphi(E_t) = \mathbb{R}$ for $t \neq 0$.

Proposition 1.7.4 *If $\varphi : E \to F$ is a morphism of vector bundles over a manifold M, then $Ker\, \varphi$ is a subbundle of E if and only if $rank\, \varphi | E_x = const$ for $x \in M$. This is also a necessary and sufficient condition for $Im\, \varphi$ to be a subbundle of F.*

Proof Set $r = rank\, E$, $s = rank\, F$.

Suppose that $Ker\, \varphi$ or $Im\, \varphi$ is a subbundle. Then $rank\, \varphi | E_x = dim\, \varphi(E_x) = const$ since $dim\, \varphi(E_x) = r - dim\, Ker\, \varphi | E_x$.

Conversely, suppose that $m = rank\, \varphi | E_x = const$. If $m = 0$, then $Ker\, \varphi = E$, $Im\, \varphi = M \times \{0\}$ and the statement is obvious. Thus, we may assume that $m \geq 1$. Let $p \in M$, and let $s_1, \ldots, s_r$ and $\sigma_1, \ldots, \sigma_s$ be frames of E and F in a neighbourhood of the point p. The matrix of the morphism φ with respect to these frames can be written in the following block-matrix form

$$\Phi = \begin{bmatrix} A_{m\times m} & B_{m\times(s-m)} \\ C_{(r-m)\times m} & D_{(r-m)\times(s-m)} \end{bmatrix}.$$

In this representation, the subscripts indicate the dimensions of the matrices making up the matrix Φ. In the case $m = s$, the matrices B and D do not enter in the block-matrix representation; if $m = r$, then C and D do not enter. The elements of the matrix Φ are smooth functions. The rank of Φ is equal to m, and rearranging the frames of E and F we have chosen, if necessary, we may assume that the matrix A at the point p is non-singular. Then this matrix is non-singular in a neighbourhood of the point p. Using the latter fact, we can find new frames $\widetilde{s}_1, \dots, \widetilde{s}_r$ and $\widetilde{\sigma}_1, \dots, \widetilde{\sigma}_s$ of E and F in a neighbourhood of p such that $\varphi(\widetilde{s}_\alpha) = \widetilde{\sigma}_\alpha$ for $\alpha = 1, \dots, m$ and $\varphi(\widetilde{s}_\beta) = 0$ for $\beta = m+1, \dots, r$. By Proposition 1.7.3, the existence of such frames would imply the claim we want to prove. So, set $\widetilde{s}_i = \sum_{j=1}^{r} x_{ij}s_j$, $i = 1, \dots, r$, and $\widetilde{\sigma}_k = \sum_{l=1}^{s} y_{kl}\sigma_l$, $k = 1, \dots, s$. We are looking for non-singular matrices $X = [x_{ij}]$ and $Y = [y_{kl}]$ whose entries are smooth functions so that, in the block-matrix representation $\widetilde{\Phi}$ of φ with respect to the new frames, the upper left block is the unit matrix and the remaining blocks are the zero matrices. Set

$$X = \begin{bmatrix} A^{-1}_{m\times m} & 0_{m\times(r-m)} \\ -(CA^{-1})_{(r-m)\times m} & I_{(r-m)\times(r-m)} \end{bmatrix}, \quad Y = \begin{bmatrix} I_{m\times m} & (A^{-1}B)_{m\times(s-m)} \\ 0_{(s-m)\times m} & -I_{(s-m)\times(s-m)} \end{bmatrix},$$

where I stands for the unit matrix. According to the standard rule for finding an inverse matrix, the elements of the matrix A^{-1} are smooth functions. So, X and Y are obviously non-singular matrices with entries smooth functions. With this choice of X and Y, we have $Y^{-1} = Y$ and

$$\widetilde{\Phi} = X\Phi Y^{-1} = \begin{bmatrix} I_{m\times m} & 0_{m\times(s-m)} \\ 0_{(r-m)\times m} & \widetilde{D}_{(r-m)\times(s-m)} \end{bmatrix}, \tag{1.13}$$

where $\widetilde{D} = CA^{-1}B - D$. Every element d of the matrix $\widetilde{D}$ is an element of a minor of order $m+1$ in $\widetilde{\Phi}$ which is a diagonal matrix of the form $diag\{1, \cdots, 1, d\}$. Since every minor of order $m+1$ in $\widetilde{\Phi}$ has zero determinant, every element of $\widetilde{D}$ is equal to zero. Now, the statement follows from Proposition 1.7.3 as we have noted. □

Example 1.7.2 Let $E \to M$ be a smooth complex vector bundle, and let J be its complex structure: $Jv = iv$ for $v \in E$ (Example 1.6.9). The endomorphism J of E can be extended by complex linearity to the complexification $E^{\mathbb{C}}$ of the real bundle E: $J(X+iY) = JX + iJY$. Since $J^2 = -Id$, the eigenvalues of the $\mathbb{C}$-linear map $J_p : E^{\mathbb{C}}_p \to E^{\mathbb{C}}_p$, $p \in M$, are $+i$ and $-i$. If $X \in E_p$, $X \neq 0$, the vector $X \mp iJ_pX$ is an eignevector of J_p corresponding to the eigenvalue $\pm i$; every eigenvector of J_p is of this type. We denote by $E^{1,0}_p$ and $E^{0,1}_p$ the eigensubspaces $E^{\mathbb{C}}_p$ corresponding to i and $-i$, respectively. Let s_1, ..., s_r be a frame of the complex bundle E. Then $s_1, Js_1, \dots s_r, Js_r$ is a frame of the real bundle E, and $s_1 - iJs_1, \dots, s_r - iJs_r$,

$s_1 + iJs_1, \dots, s_r + iJs_r$ is a frame of $E^{\mathbb{C}}$. At each point p, the first r sections of the latter frame form a basis of $E_p^{1,0}$, and the remaining sections form a basis of $E_p^{0,1}$. It follows from Proposition 1.7.3 that the subsets of $E^{\mathbb{C}}$ of eigenvectors of J corresponding to the eigenvalues i and $-i$ possess the structure of (complex) subbundles of $E^{\mathbb{C}}$. These subbundles are usually denoted by $E^{1,0}$ and $E^{0,1}$. The map $X - iJX \to X + iJX$ is an isomorphism between these two complex bundles. The map $E \ni X \to \frac{1}{2}(X - iJX) \in E^{1,0}$ is an isomorphism of the complex bundle E onto $E^{1,0}$, and the map defined by $X \to \frac{1}{2}(X + iJX)$ is an isomorphism of the conjugate bundle $\overline{E}$ onto $E^{0,1}$.

The reason for the presence of the factor $\frac{1}{2}$ is the following. Let M be a complex manifold. Consider the underlying smooth manifold M^{re} with the almost complex structure J determined by the complex structure of M. As we have seen in Example 1.2.3, the smooth complex vector bundles (TM^{re}, J) and T^hM are canonically isomorphic. Thus, $T^hM \cong (TM^{re}, J) \cong T^{1,0}M^{re}$ as smooth complex bundles where the second isomorphism is the one described above. Let $z_k = x_k + iy_k$, $k = 1, \dots, n$, be complex coordinates of M. Under the isomorphism $T^hM \cong (TM^{re}, J)$, $\dfrac{\partial}{\partial z_k} \to \dfrac{\partial}{\partial x_k}$ and $i\dfrac{\partial}{\partial z_k} \to \dfrac{\partial}{\partial y_k}$, so that $J\dfrac{\partial}{\partial x_k} = \dfrac{\partial}{\partial y_k}$. Hence, under the isomorphism $T^hM \cong T^{1,0}M^{re}$, $\dfrac{\partial}{\partial z_k} \to \dfrac{1}{2}(\dfrac{\partial}{\partial x_k} - i\dfrac{\partial}{\partial y_k})$, where these tangent vectors act in the same way on holomorphic functions.

Let $\pi_F : F \to M$ be again a subbundle of rank m of a bundle $\pi : E \to M$ of rank r. Then every fibre F_x is a subspace of the vector space E_x and we can consider the quotient space $Q_x = E_x/F_x$. Set $Q = \dot\cup_{x \in M} Q_x$, the disjoint union of the quotient spaces $Q_x = E_x/F_x$, and let $\pi_Q : Q \to M$ be the map for which $\pi_Q(Q_x) = x$. Write the vector space $\mathbb{K}^r$ as the direct sum $\mathbb{K}^r = \mathbb{K}^m \oplus \mathbb{K}^{r-m}$, and let $p_1 : \mathbb{K}^r \to \mathbb{K}^m$ and $p_2 : \mathbb{K}^r \to \mathbb{K}^{r-m}$ be the corresponding projection maps. By Proposition 1.7.1, for every point $x \in M$ there exist a neighbourhood U of x and a trivialization $\dots \varphi_U = (\pi, \Phi_U) : \pi^{-1}(U) \to U \times \mathbb{K}^r$ of E such that $\psi_U = (\pi, p_1 \circ \Phi_U)|(\pi^{-1}(U) \cap F) : \pi_F^{-1}(U) \to U \times \mathbb{K}^m$ is a trivialization of F over U. Define a map $\vartheta_U : \pi_Q^{-1}(U) = \dot\cup_{x \in U} Q_x \to U \times \mathbb{K}^{r-m}$ by $\vartheta_U([e]) = (x, \pi_2 \circ \Phi_U(e))$ for $e \in E_x$ where, as usual, $[e]$ stands for the equivalence class of e in Q_x. If $[e'] = [e]$, then $(e' - e) \in F_x$, hence $\Phi_U(e' - e) \in \mathbb{K}^m$, thus $p_2 \circ \Phi_U(e') = p_2 \circ \Phi_U(e)$. Therefore ϑ_U is a well-defined map. Clearly, this map is bijective, the inverse map being given by $(x, \eta) \to [\varphi_U^{-1}(x, \eta)]$. Moreover, for every $y \in U$, the restriction $\vartheta_U|Q_y$ is an isomorphism of the vector space Q_y onto $\{y\} \times \mathbb{K}^{r-m}$. Let φ_V be another trivialization of E with the properties stated in Proposition 1.7.1, and let $\vartheta_V : \pi_Q^{-1}(V) \to V \times \mathbb{K}^{r-m}$ be the corresponding map defined as above. If $U \cap V \neq \emptyset$, let g_{UV} be the transition function of E determined by the trivializations φ_U and φ_V, so $\varphi_U \circ \varphi_V^{-1}(x, \zeta) = (x, g_{UV}(x).\zeta))$. Then $\vartheta_U \circ \vartheta_V^{-1}(x, \eta) = (x, p_2(g_{UV}(x).\eta))$. It follows that the map $\vartheta_U \circ \vartheta_V^{-1} : (U \cap V) \times \mathbb{K}^{r-m} \to (U \cap V) \times \mathbb{K}^{r-m}$ is smooth. Now, by Proposition 1.1.1, the set Q admits a unique smooth structure such that $\pi_Q : Q \to M$ is a smooth vector bundle with trivializations $\{\vartheta_U\}$.

It follows that the transition functions of the bundle E are of the following block-matrix form

$$g_{UV}(x) = \begin{bmatrix} g^F_{UV}(x) & 0 \\ h_{UV}(x) & g^Q_{UV} \end{bmatrix}, \tag{1.14}$$

where the matrix acts on the right on the vector-rows. In this matrix, $h_{UV}(x)$ is a linear map $\mathbb{K}^{r-m} \to \mathbb{K}^r$, and the upper right block is zero since $g_{UV}(x)$ maps $\mathbb{K}^m$ into itself, as we have remarked.

Problem 1.7.1 *Define an equivalence relation on* E *by* $e' \sim e''$ *if* $\pi(e') = \pi(e'')$ *and* $(e' - e'') \in F_x$ *with* $x = \pi(e')$. *Consider the quotient topological space* $E/\sim$, *and prove that* $Q = E/\sim$ *as topological spaces.*

Problem 1.7.2 *Prove that the natural projection map* $pr : E \to Q = E/\sim$, $e \to [e]$, *is a submersion and a morphism of bundles.*

Definition 1.7.2 The bundle Q is called the quotient bundle of E by the subbundle F and is usually denoted by E/F.

Since the map $pr : E \to E/F$ is a morphism of bundles, if s is a section of E, the map $pr \circ s$ is a smooth section of E/F. Not every section of E/F arises in such a way, but the sections of the frame yielded by the trivialization ϑ_U have this form..

Example 1.7.3 Let N be a submanifold of a manifold M. Then, as we have noted above, TN can be considered in a natural way as a subbunlde of the bundle $TM|N$. The quotient bundle $(TM|N)/TN$ is called the normal bundle of the submanifold N.

A sequence of bundles over one and the same manifold M with morphisms between them

$$\cdots \longrightarrow A \stackrel{f}{\longrightarrow} B \stackrel{g}{\longrightarrow} C \longrightarrow \cdots$$

is called **exact** if $Im\ f = Ker\ g$ for all morphisms $\cdots\ f, g, \cdots$ in the sequence.

Example 1.7.4 Let F be a subbundle of a bundle E. Then the sequence

$$0 \longrightarrow F \stackrel{inclusion}{\longrightarrow} E \stackrel{projection}{\longrightarrow} E/F \longrightarrow 0 \tag{1.15}$$

is exact.

Suppose we are given a short exact sequence of bundles over a manifold M

$$0 \longrightarrow A \stackrel{f}{\longrightarrow} B \stackrel{g}{\longrightarrow} C \longrightarrow 0$$

Then, by Proposition 1.7.4, $f(A)$ is a subbunle of B since $Ker\ f|A_x = 0$ for every $x \in M$, hence $dim\ Im(f|A_x) = dim\ A_x = rank\ A = const$. The map f yields an isomorphism of the bundles A and $f(A)$ since $f|A_x$ is an isomorphism between the vector spaces A_x and $Im(f|A_x) = f(A)_x$ (Proposition 1.2.1). Note also that the quotient bundle $B/f(A)$ is isomorphic to the bundle C via the map $B/f(A) \ni [b] \to g(b) \in C$. Therefore, up to isomorphism, every short exact sequence has the form (1.15).

Problem 1.7.3 *If*

$$\cdots \longrightarrow A \stackrel{f}{\longrightarrow} B \stackrel{g}{\longrightarrow} C \longrightarrow \cdots$$

is an exact sequence of bundles over a manifold M and E is a bundle over the same manifold, then the sequence

$$\cdots \longrightarrow A \otimes E \stackrel{f \otimes Id}{\longrightarrow} B \otimes E \stackrel{g \otimes Id}{\longrightarrow} C \otimes E \longrightarrow \cdots$$

is also exact.

1.8 $\bar{\partial}$-Operator on Holomorphic Vector Bundles

Everything we have discussed so far about smooth bundles applies also to holomorphic vector bundles. For example, the transition functions of a holomorphic vector bundle with respect to holomorphic trivializations are holomorphic functions with values in $GL(\mathbb{C}^r)$. Conversely, if a covering $\{U_\alpha\}$ of a complex manifold M together with holomorphic functions $g_{\alpha\beta} : U_\alpha \cap U_\beta \to GL(\mathbb{C}^r)$ satisfying the cocycle condition are given, then there exists a holomorphic vector bundle with transition functions $g_{\alpha\beta}$, unique up to a (holomorphic) isomorphism.

Of course, there are also important differences between smooth and holomorphic bundles. We note some of them here.

1. Every smooth bundle possesses infinitely many non-zero global sections (they can vanish at some points). To construct such a section it suffices, for example, to fix a point p of the base manifold, take a non-vanishing smooth section in a neighbourhood of p (such a section always exists) and extend this local section to a global one. More precisely, we can construct a global section which coincides with the local one in a neighbourhood of p (see the proof of Lemma 2.1.1 in Chap. 2). In contrast to smooth bundles, there are holomorphic bundles which do not possess non-zero global holomorphic sections. For example, the so-called tautological bundle over a complex projective space has no non-zero global holomorphic sections; its tensor powers have the same property (see Chap. 3).

2. Any smooth second-countable manifold admits so many functionally independent smooth functions, i.e., sections of the trivial bundle, that the manifold can be embedded as a closed submanifold of $\mathbb{R}^{2n}$, $n =$ the dimension of the manifold, the

Whitney embedding theorem, see Narasimhan [20], Chap. 2, § 2.15, Corollary 2.15.11 and Remarks 2.15.12 or Lee [15], Sect. 6, Par. The Whitney Embedding Theorem. However, there are no non-constant global holomorphic functions on any connected compact complex manifold by the maximum principle. Therefore compact complex manifolds cannot be embedded in $\mathbb{C}^m$. The non-compact complex manifolds that admit an embedding as a closed complex submanifold of some $\mathbb{C}^m$ are called Stein manifolds. They can be characterized by various other properties and play a very important role in the theory of several complex variables.

3. In general, there is no natural definition of exterior differentiation on the space of sections of a smooth vector bundle. In the case of a holomorphic vector bundle $E \to M$, however, one can correctly define $\overline{\partial}$-operator on the smooth forms on M with values in E. Before giving the definition of this operator, we recall the definitions of the operators ∂ and $\overline{\partial}$ acting on the space $A^{p,q}$ of complex-valued (p,q)-forms on an almost complex manifold M.

Let (M, J) be an almost complex manifold. Consider the tangent bundle TM as a smooth complex vector bundle with complex structure J (Example 1.7.2). The sections of the bundles $T^{1,0}M$ and $T^{0,1}M$ are called vector fields of type $(1,0)$ and $(0,1)$, respectively. The sections of their dual bundles $T_{1,0}M = (T^{1,0}M)^*$ and $T_{0,1}M = (T^{0,1}M)^*$ are called 1-forms of type $(1,0)$ and $(0,1)$. Any $(1,0)$-form can be considered as a 1-form on $T^{\mathbb{C}}M$ with complex values (equivalently, as a section of $(T^{\mathbb{C}}M)^*$) that vanishes on $T^{0,1}M$; similarly $(0,1)$-forms are 1-forms on $T^{\mathbb{C}}M$ vanishing on $T^{1,0}M$. The sections of the bundles $\Lambda^p T_{1,0}M$ and $\Lambda^q T_{0,1}M$ are called forms of type $(p,0)$ and $(0,q)$, respectively. Any form of type $(p,0)$ locally has the representation $\sum_{1\le i_1<\ldots<i_p\le n} f_{i_1\ldots i_p}\alpha_{i_1}\wedge\ldots\wedge\alpha_{i_p}$, where $f_{i_1\ldots i_p}$ are smooth functions and $\alpha_1,\ldots,\alpha_n$ are $(1,0)$-forms. Therefore every $(p,0)$-form can be considered as a p-form α on $T^{\mathbb{C}}M$ (section of $\Lambda^p(T^{\mathbb{C}}M)^*$) with the property that $\alpha(v_1, \ldots, v_p) = 0$ if one of the vectors $v_1, \ldots, v_p \in T^{\mathbb{C}}M$ is of type $(0,1)$. Similarly, any $(0,q)$-form is a q-form β of $T^{\mathbb{C}}M$ such that $\beta(w_1, \ldots, w_q) = 0$ when at least one of the vectors $w_1, \ldots, w_q$ is of type $(1,0)$.

There is a natural isomorphism

$$\bigoplus_{p+q=m} (\Lambda^p T_{1,0}M \otimes \Lambda^q T_{0,1}M) \cong \Lambda^m(T_{1,0}M \oplus T_{0,1}M) = \Lambda^m(T^{\mathbb{C}}M)^* \quad (1.16)$$

such that

$$\alpha_1 \wedge \ldots \wedge \alpha_p \otimes \beta_1 \wedge \ldots \wedge \beta_q \to \alpha_1 \wedge \ldots \wedge \alpha_p \wedge \beta_1 \wedge \ldots \wedge \beta_q, \quad (1.17)$$

where $\alpha_1, \ldots, \alpha_p$ are $(1,0)$-forms and $\beta_1, \ldots, \beta_q$ are $(0,1)$-forms. The bundle $\Lambda^p T_{1,0}M \otimes \Lambda^q T_{0,1}M$ is denoted by $\Lambda^{p,q}T^*M$. Its smooth sections, viewed as $(p+q)$ - forms on $T^{\mathbb{C}}M$ under the isomorphism (1.16), are called **forms of type** (p,q). A real-valued form is said to be of type (p,q) if its complex-multilinear extension to a $(p+q)$-form on $T^{\mathbb{C}}M \times \ldots \times T^{\mathbb{C}}M$ is of type (p,q). By the correspondence (1.17), the (p,q)-forms are complex-valued $(p+q)$-forms ω on $T^{\mathbb{C}}M$ with the property that $\omega(v_1, \ldots, v_{p+q}) = 0$ whenever more than p of the vectors $v_1, \ldots, v_{p+q}$ are of type $(1,0)$, or when more than q of them are of type $(0,1)$.

If M is a complex manifold of dimension n, the holomorphic line bundle $\Lambda^n(T^hM)^*$, the determinant bundle of the cotangent bundle, is called **canonical**. Its sections are the holomorphic n-forms on M. The canonical bundle considered as a smooth complex bundle is isomorphic to $\Lambda^n T^{1,0}M$ since $T^hM \cong T^{1,0}M$ as smooth complex vector bundles. The dual bundle $\Lambda^n T^hM$ of the canonical bundle is called **anti-canonical**. The canonical and anti-canonical bundles play an important role in complex geometry.

Problem 1.8.1 *Show that an almost complex structure J is integrable (i.e., has vanishing Nijenhuis tensor) if and only if the Lie bracket of two vector fields of type $(1,0)$ with respect to J is a vector field again of type $(1,0)$.*

Problem 1.8.2 *Let ω be a real-valued 2-form on an almost complex manifold with almost complex structure J. Show that:*
(a) ω is of type $(1,1)$ exactly when $\omega(JX, JY) = \omega(X, Y)$ for every (real) tangent vectors X and Y.
(b) ω is the sum of a form of type $(2,0)$ and a form of type $(0,2)$ exactly when $\omega(JX, JY) = -\omega(X, Y)$.

Now, let M be a complex manifold. Denote its complex structure by J. If $z_1, \ldots, z_n$ are complex coordinates of M and $z_k = x_k + iy_k$, then $J\dfrac{\partial}{\partial x_k} = \dfrac{\partial}{\partial y_k}$ and $J\dfrac{\partial}{\partial y_k} = -\dfrac{\partial}{\partial x_k}$. Hence

$$\frac{\partial}{\partial z_k} = \frac{1}{2}\left(\frac{\partial}{\partial x_k} - i\frac{\partial}{\partial y_k}\right) \text{ and } \frac{\partial}{\partial \bar{z}_k} = \frac{1}{2}\left(\frac{\partial}{\partial x_k} + i\frac{\partial}{\partial y_k}\right), \quad k = 1, \ldots, n,$$

are vector fields of type $(1,0)$ and $(0,1)$, respectively, which at each point $x \in M$ form a basis of $T_x^{1,0}M$, respectively $T_x^{0,1}M$. Similarly, the forms

$$dz_k = dx_k + idy_k \text{ and } d\bar{z}_k = dx_k - idy_k$$

are of type $(1,0)$ and $(0,1)$; the forms $\{dz_k\}$ form a basis dual to $\left\{\dfrac{\partial}{\partial z_k}\right\}$ and $\{d\bar{z}_k\}$ is a basis dual to $\left\{\dfrac{\partial}{\partial \bar{z}_k}\right\}$ at each point of M. Note also that every (p,q)-form ω locally has the representation

$$\omega = \sum_{\substack{1 \le i_1 < \ldots < i_p \le n \\ 1 \le j_1 < \ldots < j_q \le n}} f_{i_1 \ldots i_p j_1 \ldots j_q} \mathrm{d}z_{i_1} \wedge \ldots \wedge \mathrm{d}z_{i_p} \wedge \mathrm{d}\bar{z}_{j_1} \wedge \ldots \wedge \mathrm{d}\bar{z}_{j_q},$$

where $f_{i_1 \ldots i_p j_1 \ldots j_q}$ are smooth functions. This representation is valid only on complex manifolds.

Given an arbitrary almost complex manifold (M, J), denote the space of (p, q)-forms on M by $A^{p,q} M$. Then the following inclusion holds for the operator d of exterior differentiation on these forms:

$$dA^{p,q} \subset A^{p+2,q-1} + A^{p+1,q} + A^{p,q+1} + A^{p-1,q+2}$$

(Kobayashi and Nomizu [14], vol. II, Chap. IX, Proposition 2.7). For $\omega \in A^{p,q}$, set

$$\partial\omega = \text{the } A^{p+1,q} - \text{component of } d\omega, \quad \overline{\partial}\omega = \text{the } A^{p,q+1} - \text{component of } d\omega.$$

If the almost complex structure J is complex, i.e., M is a complex manifold, we have

$$dA^{p,q} \subset A^{p+1,q} + A^{p,q+1}$$

(Kobayashi and Nomizu [14], vol. II, Chap. IX, Proposition 2.8). Hence, in this case, $d = \partial + \overline{\partial}$ and identity $d^2 = 0$ implies $\partial^2 = 0$, $\overline{\partial}^2 = 0$, and $\partial\overline{\partial} + \overline{\partial}\partial = 0$.

By the Cauchy-Riemann equations, a smooth complex-valued function f on a complex manifold is holomorphic exactly when $\overline{\partial} f = 0$ (equivalently, df is a $\mathbb{C}$-linear form).

Problem 1.8.3 *Prove that if* $\omega_1 \in A^{p,q}$ *and* $\omega_2 \in A^{r,s}$, *then*

$$\overline{\partial}(\omega_1 \wedge w_2) = \overline{\partial}\omega_1 \wedge \omega_2 + (-1)^{p+q}\omega_1 \wedge \overline{\partial}\omega_2.$$

Let $A^{p,q}(E)$ be the space of (p, q)-forms on M with values in a vector bundle E, i.e., the space of smooth sections σ of the bundle $\Lambda^{p,q}T^*M \otimes E$. This bundle will be denoted by $\Lambda^{p,q}(E)$. By definition, $\Lambda^{0,0} = E$ and $A^{0,0}(E)$ is the space of smooth sections of E.

If $\sigma \in A^{p,q}(E)$, then, for every $x \in M$, $\sigma(x)$ is a $\mathbb{C}$-multilinear skew-symmetric map $T_x^{1,0}M \times \ldots \times T_x^{1,0}M \times T_x^{0,1}M \times \ldots \times T_x^{0,1}M \to E_x$ smoothly depending on x, i.e., such that if $Z_1, \ldots, Z_p$ are (1,0)-smooth vector fields and $W_1, \ldots, W_q$ are (0,1)-vector fields on an open subset U of M, $\sigma(Z_1, \ldots, Z_p, W_1, \ldots, W_q)$ is a smooth map of U into E. Similarly to the case of $\mathbb{C}$-valued (p, q)-forms, the form σ can be considered as a $(p + q)$-form with values in E, i.e., a smooth section of $\Lambda^{p+q}(T^{\mathbb{C}}M)^* \otimes E$, with the property that $\sigma(v_1, ..., v_{p+q}) = 0$ if more than p of the vectors $v_1, ..., v_{p+q}$ are of type $(1, 0)$, or more than q of them are of type $(0, 1)$.

Let $\{e_1, \ldots, e_r\}$ be a local holomorphis frame of E on an open subset U of M. Then σ can be uniquely written as

$$\sigma = \sum_{i=1}^{r} \omega_i \otimes e_i,$$

where ω_i are complex values (p,q)-forms on U. Set

$$\overline{\partial}\sigma = \sum_{i=1}^{r} \overline{\partial}\omega_i \otimes e_i.$$

If $\{e'_1, \ldots, e'_r\}$ is another holomorphic frame on an open subset U' of M, then, on $U \cap U'$, we have $e'_i = \sum_{j=1}^r g_{ij} e_j$, where g_{ij} are holomorphic functions. If $\sigma = \sum_{i=1}^r \omega'_i \otimes e'_i$, then

$$\sigma = \sum_{j=1}^r \left(\sum_{i=1}^r \omega'_i . g_{ij} \right) \otimes e_j,$$

hence

$$\begin{aligned}\sum_{j=1}^r \overline{\partial}\omega_j \otimes e_j &= \sum_j \sum_i \overline{\partial}\omega'_i . g_{ij} \otimes e_j + \sum_j \left(\sum_i (-1)^{p+q} \omega'_i \wedge \overline{\partial} g_{ij} \right) \otimes e_j \\ &= \sum_{i=1}^r \overline{\partial}\omega'_i \otimes e'_i\end{aligned}$$

since $\overline{\partial} g_{ij} = 0$. Hence, the definition of $\overline{\partial}\sigma$ does not depend on the choice of the holomorphic frame $\{e_1, \ldots, e_r\}$ and we have a well-defined operator

$$\overline{\partial}_E : A^{p,q}(E) \to A^{p,q+1}(E).$$

Obviously, a smooth section s of E, i.e., an element of $\Lambda^{0,0}(E)$, is holomorphic exactly when $\overline{\partial}_E s = 0$. It is clear also that $\overline{\partial}_E^2 = 0$.

Example 1.8.1 Let M be a complex manifold and $T^h M$ its holomorphic tangent bundle. Then $E = \Lambda^p (T^h M)^*$ is a holomorphic vector bundle over M. If φ is a smooth section of E, then $\overline{\partial}_E \varphi$ is a section of the bundle $(T^{0,1}M)^* \otimes E$. As a smooth complex vector bundle $T^h M$ is isomorphic to the bundle $T^{1,0}M$. This isomorphism yields an isomorphism of $E = \Lambda^p (T^h M)^*$ considered as a smooth complex bundle with the bundle $\Lambda^p (T^{1,0}M)^*$. Hence, φ can be considered as a (p,0)-form on M. Then $\overline{\partial}\varphi$ is a $(p,1)$-form on M which can be considered as a section of the bundle $(T^{0,1}M)^* \otimes \Lambda^p (T^{1,0}M)^*$ via the isomorphism (1.16). The isomorphism $\Lambda^p (T^{1,0}M)^* \cong E$ yields an isomorphism $(T^{0,1}M)^* \otimes \Lambda^p (T^{1,0}M)^* \cong (T^{0,1}M)^* \otimes E$ in an obvious way. Thus, on one hand, we have the section $\overline{\partial}\varphi$ of the bundle $(T^{0,1}M)^* \otimes \Lambda^p (T^{1,0}M)^*$, on the other hand, the section $\overline{\partial}_E \varphi$ of the bundle $(T^{0,1}M)^* \otimes E$, and these bundles are naturally isomorphic. We claim that, under this isomorphism, $\overline{\partial}\varphi$ goes to $\overline{\partial}_E \varphi$. Indeed, let $w_1, \ldots, w_n$ be a holomorphic frame of $(T^h M)^*$. Then $w_{i_1} \wedge \ldots \wedge w_{i_p}$, $1 \leq i_1 < \ldots < i_p \leq n$, is a holomorphic frame of E, thus φ has the representation

$$\varphi = \sum f_{i_1 \ldots i_p} w_{i_1} \wedge \ldots \wedge w_{i_p}. \tag{1.18}$$

If we consider $w_1, \ldots, w_n$ as $(1,0)$-forms via the isomorphism $(T^h M)^* \cong (T^{1,0}M)^*$, then φ considered as a $(p,0)$-form has the representation (1.18). Hence, $\overline{\partial}_E \varphi = \sum \overline{\partial} f_{i_1 \ldots i_p} \otimes w_{i_1} \wedge \ldots \wedge w_{i_p}$ and $\overline{\partial}\varphi = \sum \overline{\partial} f_{i_1 \ldots i_p} \wedge w_{i_1} \wedge \ldots \wedge w_{i_p}$. Under the isomorphism $(T^{0,1}M)^* \otimes E \cong (T^{0,1}M)^* \otimes \Lambda^p (T^{1,0}M)^*$, the section $\overline{\partial}_E \varphi$ of $(T^{0,1}M)^* \otimes E$ goes to the section $\sum \overline{\partial} f_{i_1 \ldots i_p} \otimes w_{i_1} \wedge \ldots \wedge w_{i_p}$ of

$(T^{0,1}M)^* \otimes \Lambda^p(T^{1,0}M)^*$, and the latter section corresponds to $\overline{\partial}\varphi$ under the isomorphism (1.16).

Let $E \to M$ be a smooth complex vector bundle over a complex manifold M. Let $D : A^{p,q}(E) \to A^{p,q+1}(E)$ be a linear operator satisfying the Leibniz rule $D(fs) = \overline{\partial} f \otimes s + f Ds$ where f is a smooth function on M and s is a smooth section of E. We call such an operator a **Dolbeault operator** (on $A^{p,q}(E)$).

Example 1.8.2 If E is a holomorphic bundle over M, $\overline{\partial}_E$ is a Dolbeault operator.

Every Dolbeault operator $D : A^{0,0}(E) \to A^{0,1}$ can be extended to a Dolbeault operator $D : A^{p,q}(E) \to A^{p,q+1}(E)$ as follows. Take a frame $e_1, \dots, e_r$ of E. Then every $\sigma \in A^{p,q}(E)$ is represented as $\sigma = \sum_{i=1}^r \omega_i \otimes e_i$ where ω_i are (p,q)-forms on M.

If ω is a k-form on M and φ is a l-form with values in E, the $(k+l)$-form $\omega\wedge\varphi$ with values in E is defined as the anti-symmetrization of the form $(\omega\otimes\varphi)(X_1, \dots X_{k+l}) = \omega(X_1, \dots, X_k)\varphi(X_{k+1}, \dots, X_{k+l})$ for $X_i \in T_pM$.

With this definition in mind, we set

$$D\sigma = \sum_{i=1}^{r} (\overline{\partial}\omega_i \otimes e_i + (-1)^{p+q}\omega_i \wedge De_i). \tag{1.19}$$

Problem 1.8.4 *Show that the right-hand side of (1.19) does not depend on the choice of the frame $e_1, \dots, e_r$.*

Problem 1.8.5 *Prove that if $\alpha \in \Lambda^{p,q}$ and $\sigma \in \Lambda^{s,t}(E)$, then*

$$D(\alpha \wedge \sigma) = \overline{\partial}\alpha \wedge \sigma + (-1)^{p+q}\alpha \wedge D\sigma,$$

It follows from Problems 1.8.4 and 1.8.5 that identity (1.19) correctly defines a Dolbeault operator $D : A^{p,q}(E) \to A^{p,q+1}$.

As we have noted, the Dolbeault operator $\overline{\partial}_E$ of a holomorphic vector bundle E has the property that $\overline{\partial}_E^2 = 0$. The identity $D^2 = 0$ for a Dolbeault operator D characterizes the holomorphic bundles among smooth ones. This is the contents of the following result known as the Koszul-Malgrange or Grothendieck-Koszul-Malgrange theorem.

Proposition 1.8.1 *If a smooth complex vector bundle $E \to M$ over a complex manifold M admits a Dolbeault operator D satisfying $D^2 = 0$, then E admits a unique structure of a complex manifold rendering $E \to M$ to a holomorphic vector bundle such that $\overline{\partial}_E = D$.*

Problem 1.8.6 *Prove the uniqueness assertion.*

We refer to Moroianu [18], Part 1, Sect. 2, Theorem 3.2 for a proof of the existence part of the proposition.

Connections on Vector Bundles 2

2.1 Definition of a Connection. Matrix of a Connection with Respect to a Frame

Let $\pi : E \to M$ be a smooth $\mathbb{K}$-vector bundle. Denote by $A^p(E)$ the space of sections of the bundle $\Lambda^p T^*M \otimes E$ (where $A^0(E)$ is the space of the sections of E), i.e., $A^p(E)$ is the space of p-forms on M with values in the bundle E. The ring of smooth functions on M will be denoted by $\mathcal{F}(M)$, and the space of vector fields on M by $\chi(M)$; $\chi(M)$ is a module over $\mathcal{F}(M)$.

Definition 2.1.1 A connection on E is a $\mathbb{K}$-linear map

$$D : A^0(E) \to A^1(E),$$

satisfying the Leibniz rule

$$D(fs) = df \otimes s + f Ds, \quad s \in A^0(E),\ f \in \mathcal{F}(M).$$

In the case when E is the tangent bundle of a smooth manifold, the connections on E are usually called connections on the manifold.

Let D be a connection on a bundle E. For $v \in T_pM$, the element $Ds(v) \in E_p$ will be denoted by $D_v s$; it is called **the covariant derivative** of s in direction v. For each vector field X on M, we denote by $D_X s$ the section of E defined by $p \to D_{X_p} s$. Note that $D_X s$ is linear with respect to X over the ring $\mathcal{F}(M)$ of smooth functions on M, while $D_X s$ satisfies the Leibniz rule with respect to s which can be written as $D_X(fs) = X(f)s + f D_X s$.

If E is a complex vector bundle and $Z = X + iY$, $X, Y \in TM$, is a complex tangent vector of M, we can define $D_Z s$ setting $D_Z s = D_X s + i D_Y s$.

J. Davidov, *Vector Bundles and Connections*, Compact Textbooks in Mathematics,
https://doi.org/10.1007/978-3-032-07403-4_2

So, a connection is essentially a way of differentiating smooth sections. However, this is only an approximation of differentiation because the mixed derivatives $D_X D_Y s$ and $D_Y D_X s$ are not always equal.

Lemma 2.1.1 *Let U be an open subset of M, $s \in A^0(M)$ and $s|U = 0$. Then $Ds|U = 0$.*

Proof Let $p \in U$, and let $V \subset U$ be a coordinate neighbourhood of the point p. Then there exists a smooth function φ on V such that $supp\,\varphi$ is a compact subset of V and $\varphi \equiv 1$ in a neighbourhood of p (for the existence of such a function see, e.g., Brickell and Clark [3], Chap. 3, Lemma 3.4.2 or Conlon [4], Chap. 2, Theorem 2.6.1; recall that the support $supp\, f$ of a function f defined on a topological space A is the closure of the set $\{x \in A : f(x) \neq 0\}$). Set

$$\widetilde{\varphi} = \begin{cases} \varphi \text{ on } & V \\ 0 \text{ on } & M \setminus V. \end{cases}$$

The function $\widetilde{\varphi}$ is smooth on M since it is smooth on the open sets V and $M \setminus supp\,\varphi$ whose union is M. Since $s|U = 0$, the identity $s = (1 - \widetilde{\varphi})s$ holds. Hence for $v \in T_pM$, $D_v s = v(1 - \widetilde{\varphi})s(p) + (1 - \widetilde{\varphi})(p) D_v s = 0$, since $s(p) = 0$ and $\widetilde{\varphi}(p) = 1$. □

It follows from Lemma 2.1.1 that if U is an open subset of M, then we can define a connection on the bundle $E|U$ as follows. Let σ be a section of the bundle $E|U$. For every point $p \in U$, there exists a globally defined section $\widetilde{\sigma}$ of E which coincides with σ in a neighbourhood of the point p. For, let φ be a function as in the proof of Lemma 2.1.1. Then

$$\widetilde{\sigma} = \begin{cases} \varphi\sigma \text{ on } & V \\ 0 \text{ on } & M \setminus V \end{cases}$$

is a global smooth section of E which coincides with σ in a neighbourhood of the point p. For $v \in T_pM$, set $D_v^U \sigma = D_v \widetilde{\sigma}$. By Lemma 2.1.1, $D_v \widetilde{\sigma}$ does not depend on the choice of the extension $\widetilde{\sigma}$ of σ. Thus, for every global section ξ of E, we have $D^U(\xi|U) = (D\xi)|U$. Lemma 2.1.1 implies that there is a unique connection on the bundle $E|U$ with this property. Indeed, let ∇ be such a connection on $E|U$, and let σ be a section of the bundle $E|U$. For every $p \in U$, we can construct a global section $\widetilde{\sigma}$ of E which coincides with σ in a neighbourhood of p. Then by Lemma 2.1.1, at the point p, $\nabla\sigma = \nabla(\widetilde{\sigma}|U) = (D\widetilde{\sigma})|U = D^U\sigma$. Further on, the connection D^U will simply be denoted by D when the set U is understood.

Remark 2.1.1 The consideration above shows that for every $a \in E$ there exists a global section s such that $s(\pi(a)) = a$. Indeed, let $s_1, \ldots, s_r$ be a frame of E in a neighbourhood of the point $p = \pi(a)$, and let $a = \sum_{i=1}^r \lambda_i s_i(p)$, $\lambda_i \in \mathbb{K}$. Then $\sigma = \sum_{i=1}^r \lambda_i s_i$ is a local section with $\sigma(p) = a$. This section can be extended to a global section s which coincides with σ in a neighbourhood of the point p.

Problem 2.1.1 Check that D^U is really a connection on the bundle $E|U$.

Example 2.1.1 Let $E = M \times \mathbb{K}^r$ be the trivial bundle over M. Any section of this bundle is of the form $s(x) = (x, f(x))$, $x \in M$, where $f : M \to \mathbb{K}^r$ is a smooth vector-valued function. Setting $(Ds)(x) = (x, (df)(x))$ defines a connection on E. This connection is called trivial and is denoted by d.

Now, we recall the notion of a partition of unity which is often used to glue together locally defined objects into a global one.

A family $\{\rho_\alpha\}_{\alpha\in A}$ of globally defined smooth functions on a manifold M is called a partition of unity if (1) $0 \leq \rho_\alpha \leq 1$; (2) the supports $supp\ \rho_\alpha$ constitute a locally finite family; (3) $\sum_{\alpha\in A} \rho_\alpha(x) = 1$ for every $x \in M$.

The condition (2) means that every point of M admits a neighbourhood that intersects only a finite number of the sets $supp\ \rho_\alpha$. This implies that the sum in (3) is well defined in such a neighbourhood (only a finite number of the summands in (3) are not zero).

A partition of unity $\{\rho_\alpha\}_{\alpha\in A}$ is said to be subordinated to an open cover $\{U_\alpha\}_{\alpha\in A}$ of M if $supp\ \rho_\alpha \subset U_\alpha$ for every $\alpha \in A$.

An important theorem in manifold theory states that for every open cover of a paracompact smooth manifold there exists a subordinated partition of unity. Recall that a topological space X is said to be paracompact if it is Hausdorff and for every open cover $\{U_\alpha\}$ of X there exists an open cover $\{V_i\}$ of X such that $V_i \subset$ of some U_α and which is locally finite (every point of X has a neighbourhood intersecting only a finite number of the sets V_i); it is said that a cover $\{V_i\}$ is a refinement of a cover $\{U_\alpha\}$ if every V_i is a subset of some U_α.

A useful criterion for paracompactness states that every locally compact Hausdorff space with countable base is paracompact, see, e.g., Conlon [4], Chap. 1, Theorem 1.4.5. Note also that for a locally Euclidean, Hausdorff, connected space the following three conditions are equivalent. (1) X is sigma compact, i.e., X is the union of countably many compact subspaces; (2) X admits a countable base; (3) X is paracompact; for a proof see the website https://ncatlab.org/nlab/show/topological+manifold.

A proof of the existence of a subordinated partition of unity on a paracompact manifold can be seen, e.g., in Conlon [4], Chap. 3, Theorem 3.5.4. The proof therein is a modification of the proof of Theorem 1.4.11 where, in order to shorten some arguments, the condition that the space is regular is included in the definition of a paracompact space. Thus, the proof of Theorem 3.5.4. makes use of the claim that every paracompact space is regular, i.e., every point and every non-containing it closed set possess non-intersecting open neighbourhoods. In fact, one can prove that every paracompact space (in the sense used here) is normal, i.e., every two non-intersecting closed sets possess non-intersecting open neighbourhoods, Munkres [19], Chap. 6, Sect. 41, Theorem 41.1.

For another proof of the existence theorem, see the website https://ncatlab.org/nlab/show/partition+of+unity.

We refer to Brickell and Clark [3], Chap. 3, Sect. 3.4 and Warner [24], Chap. 1, Theorem 1.11. for a version of the existence theorem in which the supports of the functions ρ_α are compact.

Finally, note that, on a real analytic or a complex manifold, there is no partition of unity with real analytic or holomorphic functions since every such function with compact support on a connected manifold is constant.

Using partition of unity, it is easy to show that every smooth vector bundle $\pi : E \to M$ over a paracompact manifold M possesses a connection. Indeed, let $\{U_\alpha\}$ be an open cover of M such that E is trivial on every U_α with trivializations $\varphi_\alpha : \pi^{-1}(U_\alpha) \to U_\alpha \times \mathbb{K}^r$. Making use of the isomorphism φ_α, we can transfer the trivial connection on $U_\alpha \times \mathbb{K}^r$ to the bundle $E|U_\alpha = \pi^{-1}(U_\alpha)$. Thus, we obtain a connection on $E|U_\alpha$. Let $\{\rho_\alpha\}$ be a partition of unity on M subordinated to the cover $\{U_\alpha\}$. For every smooth section s of E, set $Ds = \sum_\alpha \rho_\alpha D_\alpha(s|U_\alpha)$. Then D is a connection on E.

Let $\pi : E \to M$ be a smooth vector bundle and D a connection on it. Let $e = (e_1, ..., e_r)$ be a frame of E on an open set $U \subset M$. The covariant derivatives of the sections e_i have the form

$$De_i = \sum_{j=1}^{r} \theta_{ij} \otimes e_j, \quad i = 1, ..., r,$$

where θ_{ij} are $\mathbb{K}$-valued 1-forms on U. The matrix $\theta_e = [\theta_{ij}]$ is called the **matrix of the connection** D with respect to the frame e. This matrix determines the connection on U. For, let $\sigma : U \to E$ be a section and $\sigma = \sum_{i=1}^{r} \sigma_i e_i$; here σ_i are smooth functions on U. Then by the Leibniz rule

$$D\sigma = \sum_i d\sigma_i \otimes e_i + \sum_{i,j} \sigma_i \theta_{ij} \otimes e_j = \sum_j (d\sigma_j + \sum_i \sigma_i \theta_{ij}) \otimes e_j. \tag{2.1}$$

Example 2.1.2 Let $a_1, ..., a_r$ be a basis of $\mathbb{K}^r$ and let $e_i(x) = (x, a_i), i = 1, ..., r$, $x \in M$, be the corresponding frame of the trivial bundle $M \times \mathbb{K}^r$. Then the matrix of the trivial connection d on the trivial bundle with respect to the frame $e = (e_1, ..., e_r)$ is $\theta_e = 0$.

Now, let $e' = (e'_1, ..., e'_r)$ be another frame of E on U and $[\theta'_{ij}]$ the matrix of the connection D with respect to e'. We have $e'_i = \sum_{j=1}^{r} g_{ij} e_j$, where g_{ij} are smooth functions. Then

$$\sum_j \theta'_{ij} \otimes e'_j = De'_i = \sum_j dg_{ij} \otimes e_j + \sum_{j,k} g_{ij}\theta_{jk} \otimes e_k = \sum_k (dg_{ik} + g_{ij}\,\theta_{jk}) \otimes e_k.$$

Therefore, if we denote the matrix $[g_{ij}]$ by g, then

$$\theta_{e'} = dg.g^{-1} + g.\theta_e.g^{-1}. \tag{2.2}$$

Conversely, suppose we are given a family of frames $\{e\}$ of E on open sets $\{U\}$ constituting a cover of M and matrices θ_e of 1-forms on U such that on $U \cap U' \neq \emptyset$ the identity (2.2) relating the corresponding matrices under change of the frame holds. Then the right-hand side of identity (2.1) correctly defines a connection on E.

Let E be a smooth complex vector bundle. For $x \in M$, denote the complex structure of the fibre E_x by J_x ; $x \to J_x$ is a smooth section of the bundle $Hom(E, E)$. If D is a connection on E, then D is $\mathbb{C}$-linear, i.e., for every section s of E the identity $DJs = JDs$ holds. Conversely, let D be a connection on the bundle E considered as a smooth real vector bundle such that $DJs = JDs$. Then D is a $\mathbb{C}$-linear map, hence it is a connection on the complex vector bundle E.

We can extend the connection D on E by complex linearity to a connection on the complexification $E^{\mathbb{C}}$ of E. The connection $D^{\mathbb{C}}$ on $E^{\mathbb{C}}$ defined in this way can also be obtained as follows. Let $E^{\mathbb{C}} = E^{1,0} \oplus E^{0,1}$ be the decomposition of $E^{\mathbb{C}}$ into $(1, 0)$ and $(0, 1)$-subbundles. The complex bundles E and $E^{1,0}$ are isomorphic under the map $E \ni \xi \to \frac{1}{2}(\xi - iJ\xi) \in E^{1,0}$. Using this map, we can transfer the connection of E to a connection on $E^{1,0}$. Denote the connection on $E^{1,0}$ obtained in this way again by D. This connection is naturally extended to $E^{\mathbb{C}}$: for any $Z \in TM$ and a section ψ of $E^{0,1}$, set $D_Z\psi = \overline{(D_{\overline{Z}}\overline{\psi})}$. The resulting connection on $E^{\mathbb{C}}$ coincides with the connection $D^{\mathbb{C}}$.

2.2 Algebraic Operations on Vector Bundles and Induced Connections

Let $E \to M$ and $F \to M$ be smooth $\mathbb{K}$-vector bundles on a manifold M endowed with connections D^E and D^F. These connections yield in a natural way connections on the bundles obtained from E and F by algebraic operations.

1. The connections D^E and D^F induce a connection $D^{E\oplus F}$ on the bundle $E \oplus F$ defined by

$$D_X^{E\oplus F}(\xi + \eta) = D_X^E\xi + D_X^F\eta, \quad X \in TM,$$

where ξ is a section of E and η a section of F.

Let $e = (e_1, ..., e_r)$ and $f = (f_1, ..., f_s)$ be frames of E and F on an open subset U of M. Denote the corresponding matrices of the connections D^E and D^F by θ_e and θ_f. Under this notation, the matrix θ of the connection $D^{E\oplus F}$ with respect to the frame $(e_1, ..., e_r, f_1, ..., f_s)$ is $\theta_e + \theta_f$. Here θ_e stands for the $(r+s) \times (r+s)$-matrix whose upper left $r \times r$-block is the connection matrix θ_e and the other blocks are zero; similarly, θ_f denotes the $(r+s) \times (r+s)$-matrix whose lower right $s \times s$-block is the connection matrix θ_f and the other blocks are zero.

2. There is a unique connection $D^{E\otimes F}$ on the bundle $E \otimes F$ such that

$$D_X^{E\otimes F}(\xi \otimes \eta) = D_X^E\xi \otimes \eta + \xi \otimes D_X^F\eta, \quad X \in TM.$$

Suppose that a connection with this property exists and let σ be a section of $E \otimes F$. The sections $e_i \otimes f_j$, $1 \leq i \leq r$, $1 \leq j \leq s$, form a frame of the bundle $E \otimes F$, hence $\sigma = \sum_{i=1}^{r} \sum_{j=1}^{s} \sigma_{ij} e_i \otimes f_j$. Then

$$D_X^{E \otimes F} \sigma = \sum_{i=1}^{r} \sum_{j=1}^{s} [X(\sigma_{ij}) e_i \otimes f_j + \sigma_{ij} D_X^E e_i \otimes f_j + \sigma_{ij} e_i \otimes D_X^F f_j].$$

Obviously, this identity implies the uniqueness of $D^{E \otimes F}$. Conversely, it is easy to check that the right-hand side of the identity above does not depend on the choice of the frames $e_1, \dots, e_r$ and $f_1, \dots, f_s$, and defines a connection on $E \otimes F$ with the desired property.

We have

$$D^{E \otimes F}(e_i \otimes f_j) = \sum_{k=1}^{r} \sum_{l=1}^{s} (\theta_{ik}^E \delta_{jl} + \delta_{ik} \theta_{jl}^F) e_k \otimes f_l.$$

Let $I_r = [\delta_{ik}]$ and $I_s = [\delta_{jl}]$ be the unit matrices of order r and s. Then it follows from the latter identity that the matrix of the connection $D^{E \otimes F}$ with respect to the frame $e_i \otimes f_j$ is $\theta_e \otimes I_s + I_r \otimes \theta_f$ where $\otimes$ stands for the Kronecker product of matrices.

3. Any connection D^E yields a connection D^{E^*} on the dual bundle E^* such that

$$(D_X^{E^*} \alpha)(s) = X(\alpha(s)) - \alpha(D_X^E s), \tag{2.3}$$

where α and s are sections of E^* and E, respectively, and $X \in T_p M$. A key point in showing that a connection on E^* can be defined via formula (2.3) is the fact that the operator $D_X^{E^*} \alpha$ acting on the sections s of E by formula (2.3) is linear with respect to the smooth functions on M. To see this, let f be a smooth function in a neighbourhood of a point $p \in M$. Then

$$\begin{aligned}
&(D_X^{E^*} \alpha)(fs) = X(f\alpha(s)) - \alpha(X(f)s + f(p) D_X s) \\
&= X(f)\alpha(s) + f(p) X(\alpha(s)) - X(f)\alpha(s) - f(p)\alpha(D_X^E s) = f(p)(D_X^{E^*} \alpha)(s).
\end{aligned}$$

It follows that $(D_X^{E^*} \alpha)(s)$ depends only on the value of the section s at the point p. To see this, consider two sections s and σ in a neighbourhood of p such that $s(p) = \sigma(p)$. Let $e_1, \dots, e_r$ be a frame of E in a neighbourhood of p, and let $s = \sum_{i=1}^{r} f_i e_i$, $\sigma = \sum_{i=1}^{r} g_i e_i$. Then $f_i(p) = g_i(p)$, hence

$$\begin{aligned}
(D_X^{E^*} \alpha)(s) &= \sum_{i=1}^{r} (D_X^{E^*} \alpha)(f_i e_i) = \sum_{i=1}^{r} f_i(p)(D_X^{E^*} \alpha)(e_i) = \sum_{i=1}^{r} g_i(p)(D_X^{E^*} \alpha)(e_i) \\
&= (D_X^{E^*} \alpha)(\sigma).
\end{aligned}$$

Now, take $e \in E_p$. As we have seen, there is a section s of E in a neighbourhood of p such that $s(p) = e$. Since $(D_X^{E^*} \alpha)(s)$ does not depend on the choice of s, we may set $(D_X^{E^*} \alpha)(e) = X(\alpha(s)) - \alpha(D_X^E s)$. It is easy to check that this defines a connection on the bundle E^*.

4. On the bundle $Hom(E, F)$, we can define a connection by

$$(D_X\varphi)(s) = D^F_X\varphi(s) - \varphi(D^E_X s), \quad X \in TM,$$

where φ is a section of $Hom(E, F)$ and s a section of E.

Under the canonical isomorphism $Hom(E, F) \cong E^* \otimes F$, this connection goes to the connection of $E^* \otimes F$ induced by D^{E^*} and D^F.

5. There is a unique connection D on the bundle $\Lambda^p E$ such that

$$D_X(s_1\wedge...\wedge s_p) = D_X s_1\wedge s_2\wedge...\wedge s_p + s_1\wedge D_X s_2\wedge...\wedge s_p + \cdots + s_1\wedge s_2\wedge...\wedge D_X s_p;$$

where $s_1, ..., s_p$ are sections of the bundle E, and $X \in TM$.

6. There is a unique connection D on the bundle $\odot^p E$ such that

$$D_X(s_1\odot...\odot s_p) = D_X s_1\odot s_2\odot...\odot s_p + s_1\odot D_X s_2\odot...\odot s_p + \cdots + s_1\odot s_2\odot...\odot D_X s_p.$$

Problem 2.2.1 Find the matrices of the connections defined in items 3–6 with respect to the frames of the corresponding bundle yielded by frames of E and F.

Problem 2.2.2 Let D be a connection on a $\mathbb{K}$-vector bundle $E \to M$. Such a connection induces a connection on the dual bundle E^* which in turn induces a connection on the bundle $\otimes^p E^*$. Denote the latter connection again by D. Let ω be a section of $\otimes^p E^*$, so, for $x \in M$, $\omega_x : E_x \times ... \times E_x \to \mathbb{K}$ is a multilinear p-form. Show that if $s_1, ..., s_p$ are sections of E, and X is a tangent vector of M

$$(D_X\omega)(s_1, ..., s_p) = X(\omega(s_1, ..., s_p)) - \omega(D_X s_1, s_2, ..., s_p) - \cdots - \omega(s_1, s_2, ..., D_X s_p).$$

2.3 Induced Connection of a Pull-Back Bundle

Let $f : M \to N$ be a smooth map of smooth manifolds and $\pi : E \to N$ a smooth vector bundle. Let $\widetilde{\pi} : \widetilde{E} = f^*E \to M$ be the pull-back bundle of E under f, $\widetilde{E} = \{(x, e) \in M \times E : f(x) = \pi(e)\}$. If s is a section of E, then $M \ni x \to (x, f(s(x)))$ is a section of the bundle $\widetilde{E}$ which we denote by $s \circ f$. Not every section of $\widetilde{E}$ is of this form. But, if $s_1, ..., s_r$ is a frame of E, then $s_1 \circ f, ..., s_r \circ f$ is a frame of $\widetilde{E}$ and every section σ of $\widetilde{E}$ is uniquely represented as $\sigma = \sum_{i=1}^r \lambda_i(s_i \circ f)$ where λ_i are smooth functions.

If D is a connection on E, there is a unique connection $\widetilde{D}$ on the bundle $\widetilde{E}$ such that for every section s of E

$$\widetilde{D}_X(s \circ f) = D_{f_*X}s, \quad X \in TM.$$

For, if a connection $\widetilde{D}$ on $\widetilde{E}$ possesses this property and σ is a section of $\widetilde{E}$ represented in the form $\sigma = \sum_{i=1}^{r} \lambda_i (s_i \circ f)$, then

$$\widetilde{D}_X \sigma = \sum_{i=1}^{r} [X(\lambda_i) s_i(f(p)) + \lambda_i(p) D_{f_* X} s_i], \quad X \in T_p M.$$

The latter identity implies the uniqueness of $\widetilde{D}$. It is easy to check that the right-hand side of this identity does not depend on the choice of the frame $s_1, ..., s_r$ and it defines a connection on $\widetilde{E}$ with the desired property.

It is clear that if $\theta = [\theta_{ij}]$ is the matrix of the connection D with respect to the frame $s_1, ..., s_r$, then $f^*\theta = [\theta_{ij} \circ f]$ is the matrix of the connection $\widetilde{D}$ with respect to the frame $s_1 \circ f, ..., s_r \circ f$.

We will also use the notation $f^* D$ for the connection $\widetilde{D}$ on the bundle $f^* E$ induced by D.

Problem 2.3.1 If $f : M \to N$ is a constant map, the bundle $f^* E$ is (isomorphic to) the trivial bundle $M \times \mathbb{K}$. Prove that for every connection D on the bundle E, the connection $\widetilde{D} = f^* D$ is the trivial one.

Problem 2.3.2 Let $E \to N$ be a vector bundle over a manifold N and D a connection on E. Let $h : M' \to M$ and $f : M \to N$ be smooth maps. Denote by $\widetilde{D}$ the connection on the bundles $f^* E$ induced by D, and let $\widetilde{D}'$ be the connection on $h^*(f^* E)$ induced by $\widetilde{D}$. Recall that $(f \circ h)^* E \cong h^*(f^* E)$. Show that $\widetilde{D}'$ is the connection on the bundle $(f \circ h)^* E$ induced by the connection D on E.

Let $E \to N$ be a vector bundle over a manifold N and D a connection on E. Let $f : M_1 \times M_2 \to N$ be a smooth map. Denote by $\widetilde{D}$ the connection on the bundle $f^* E$ induced by D. Let $(a, b) \in M_1 \times M_2$ and let $f^b : M_1 \to N$ and $f^a : M_2 \to N$ be the maps $f^b(x) = f(x, b)$ and $f^a(y) = f(a, y)$. Denote by $\widetilde{D}^b$ and $\widetilde{D}^a$ the connections on the bundles $(f^b)^* E$ and $(f^a)^* E$ induced by D. Let $j^b : M_1 \to M_1 \times M_2$ and $j^a : M_2 \to M_1 \times M_2$ be the maps $j^b(x) = (x, b)$ and $j^a(y) = (a, y)$. If s is a section of the bundle $f^* E$ on a neighbourhood of the point (a, b), then $s^b = s \circ j^b$ and $s^a = s \circ j^a$ are sections of the bundles $(f^b)^* E$ and $(f^a)^* E$ defined on a neighbourhood of a and b, respectively. Take $(X, Y) \in T_a M_1 \times T_b M_2$. Under the canonical isomorphism $T_a M_1 \times T_b M_2 \cong T_{(a,b)}(M_1 \times M_2)$, the vector (X, Y) is identified with $Z = j^b_* X + j^a_* Y$. As usual, identify the fibre of the bundle $f^* E$ over the point (a, b) with $E_{f(a,b)}$; similarly for the fibre of $(f^b)^* E$ over the point a and the fibre of $(f^a)^* E$ over b. Then

$$\widetilde{D}_Z s = \widetilde{D}^b_X s^b + \widetilde{D}^a_Y s^a. \tag{2.4}$$

To see this, take a frame $s_1, ..., s_r$ of E in a neighbourhood of the point $f(a, b)$. Then $\widetilde{s}_i = s_i \circ f$, $i = 1, ..., r$, is a frame of $f^* E$ in a neighbourhood of the point (a, b). Similarly, $\widetilde{s}^b_i = s_i \circ f^b$ and $\widetilde{s}^a_i = s \circ f^a$ are frames of $(f^b)^* E$ and $(f^a)^* E$ in neighbourhoods of a and b, respectively. The section s has the form $s = \sum_{i=1}^{r} \varphi_i \widetilde{s}_i$

where φ_i are smooth functions on a neighbourhood of (a, b). Then $s^b = \sum_{i=1}^r (\varphi_i \circ j^b)\widetilde{s}_i^b$ and $s^a = \sum_{i=1}^r (\varphi_i \circ j^a)\widetilde{s}_i^a$. These identities imply

$$\widetilde{D}_Z s = \sum_{i=1}^r Z(\varphi_i) s_i(f(a,b)) + \sum_{i=1}^r \varphi_i(a,b) D_{f_* Z} s_i$$
$$\widetilde{D}_X^b s^b = \sum_{i=1}^r X(\varphi_i \circ j^b) s_i(f(a,b)) + \sum_{i=1}^r \varphi_i(a,b) D_{f_*^b X} s_i,$$
$$\widetilde{D}_X^a s^a = \sum_{i=1}^r Y(\varphi_i \circ j^a) s_i(f(a,b)) + \sum_{i=1}^r \varphi_i(a,b) D_{f_*^a Y} s_i.$$

Since $Z = j_*^b X + j_*^a Y$, we have $Z(\varphi_i) = X(\varphi_i \circ j^b) + Y(\varphi_i \circ j^a)$. Moreover, $f_* Z = f_*(j_*^b X) + f_*(j_*^a Y) = (f \circ j^b)_* X + (f \circ j^a)_* Y = f_*^b X + f_*^a Y$. This proves identity (2.4).

2.4 Metrics on Vector Bundles and Metric Connections

Definition 2.4.1 Let E be a real smooth vector bundle over a smooth manifold M. A metric on E is a section of the bundle $E^* \odot E^*$ such that for every $x \in M$ the symmetric bilinear form $g_x : E_x \times E_x \to \mathbb{R}$ is positive definite.

Recall that the smoothness condition for g is equivalent to the requirement that for every two smooth local sections ξ and η of E the function $x \to g_x(\xi(x), \eta(x))$ is smooth. We will also say that g is an Euclidean metric on E. In the case when E is the tangent bundle of M, we used the term "Riemannian metric". A smooth manifold with a Riemannian metric on its tangent bundle is called a Riemannian manifold.

We define a Hermitian metric h on a smooth complex vector bundle E in a similar way. For every $x \in M$, a Hermitian scalar product h_x is defined on the fibre E_x, smoothly depending on the point x. A complex bundle equipped with a Hermitian metric is called Hermitian. A complex manifold with a Hermitian metric on its tangent bundle is called a Hermitian manifold.

Proposition 2.4.1 *Every (complex) bundle E over a paracompact manifold M admits a (resp. Hermitian) metric.*

The **proof** of this fact is similar to the proof of the existence of a connection on E. For every trivializing neighbourhood $U \subset M$ of E, we can transfer on $\pi^{-1}(U)$ the metric on $U \times \mathbb{K}^r$, and then to glue the metrics on the open sets $\pi^{-1}(U)$ to a metric on E by means of a partition of unity.

If a bundle $E \to M$ is endowed with a metric g (Euclidean or Hermitian), then applying the standard Gram-Schmidt orthogonalization procedure we can obtain an orthonormal frame from any frame of E.

Problem 2.4.1 Prove that a real bundle E of rank r admits an Euclidean metric if and only if E admits an atlas of local trivializations whose transition functions take values in the orthogonal group $O(r)$. Similarly, a complex bundle admits a Hermitian metric if and only if it possesses transition functions with values in the unitary group $U(r)$.

Any transition functions $\{g_{\alpha\beta}\}$ of a bundle take values in the group $GL(\mathbb{K}^r)$. If we can find an atlas of trivializations whose transition functions take values in a Lie subgroup G of $GL(\mathbb{K}^r)$, we say that **the structure group of the bundle can be reduced to the group** G. In these terms, we can say that a bundle admits a metric if and only if its structure group can be reduced to the orthogonal (unitary) group.

Let $E \to M$ be a smooth $\mathbb{K}$-vector bundle and g a metric on E, Euclidean in the case $\mathbb{K} = \mathbb{R}$ and Hermitian if $\mathbb{K} = \mathbb{C}$. Also, let D be a connection on E.

Definition 2.4.2 It is said that D is compatible with g or that D is a metric connection if for every (local) sections ξ and η of E and every tangent vector $X \in TM$

$$X(g(\xi, \eta)) = g(D_X\xi, \eta) + g(\xi, D_X\eta). \tag{2.5}$$

This condition can also be written as $d(g(\xi, \eta)) = g(D\xi, \eta) + g(\xi, D\eta)$.

Apparently, the compatibility condition (2.5) is analogous to Leibniz's rule for differentiating the scalar product of vector-valued functions.

If g is a Hermitian metric, identity (2.5) implies that for every complex tangent vector Z of M

$$Z(g(\xi, \eta)) = g(D_Z\xi, \eta) + g(\xi, D_{\overline{Z}}\eta).$$

A connection D determines a connection on the dual bundle E^*, which in turn yields a connection on the bundle $E^* \odot E^*$. Denote the latter connection again by D. Then the condition (2.5) can be written as $Dg = 0$ (see Problem 2.2.2).

Example 2.4.1 Every metric on $\mathbb{K}^r$ defines a metric on the trivial bundle $M \times \mathbb{K}^r$ in an obvious way. The trivial connection d on $M \times \mathbb{K}^r$ is compatible with this metric.

Problem 2.4.2 Let g be a metric on a vector bundle $E \to M$. For every $x \in M$, let $\mathcal{A}(E_x, E_x)$ be the vector space of g-skew-symmetric endomorphisms of the fibre E_x.

(i) Prove that $\mathcal{A}(E, E) = \dot{\bigcup}_{x\in M} \mathcal{A}(E_x, E_x)$ admits the structure of a subbundle of the bundle $Hom(E, E)$ of endomorphsims of E.

(ii) Let D be a metric connection on E and denote the induced connection on $Hom(E, E)$ again by D. Prove that if φ is a section of the bundle $\mathcal{A}(E, E)$ considered as a section of $Hom(E, E)$ and if X is a vector field on M, then $D_X\varphi$ is a section of $\mathcal{A}(E, E)$ (thus, D induces a connection on $\mathcal{A}(E, E)$).

Suppose E is a real vector bundle with Euclidean metric g and a metric connection D. Let $e_1, ..., e_r$ be an orthonormal frame. Denote the matrix of the connection D with respect to this frame by $[\theta_{ij}]$, thus $De_i = \sum_{j=1}^r \theta_{ij} \otimes e_j$. Then $g(De_i, e_j) + g(e_i, De_j) = dg(e_i, e_j) = d\delta_{ij} = 0$. Hence

$$\theta_{ij} + \theta_{ji} = 0,$$

i.e., the matrix of a metric connection with respect to an orthonormal frame is skew-symmetric. Conversely, suppose that a connection D is given on E such that for every point p of M there exists an orthonormal frame $e_1, ..., e_r$ in a neighbourhood of p for which the connection matrix of D is skew-symmetric. Let $\xi = \sum_{i=1}^r \lambda_i e_i$ and $\eta = \sum_{j=1}^r \mu_j e_j$ be smooth sections of E. Then

$$\begin{aligned} g(D\xi, \eta) + g(\xi, D\eta) &= \sum_{i,j}[g(d\lambda_i \otimes e_i + \lambda_i De_i, \mu_j e_j) + g(\lambda_i e_i, d\mu_j \otimes e_j + \mu_j De_j)] \\ &= \sum_i [d\lambda_i \cdot \mu_i + \lambda_i d\mu_i] + \sum_{i,j}[\lambda_i \mu_j \theta_{ij} + \lambda_i \mu_j \theta_{ji}] = \sum_i [d\lambda_i \cdot \mu_i + \lambda_i d\mu_i] = d\sum_i \lambda_i \mu_i \\ &= dg(\xi, \eta). \end{aligned}$$

Therefore the connection D is metric.

Now, let E be a complex vector bundle with a Hermitian metric g and a metric connection D. If $e_1, ..., e_r$ is an orthonormal (unitary) frame and $Z \in T_x^{\mathbb{C}} M$, then

$$\begin{aligned} 0 = Z(\delta_{ij}) &= Z(g(e_i, e_j)) = g(D_Z e_i, e_j) + g(e_i, D_{\overline{Z}} e_j) = \theta_{ij}(Z) + \overline{\theta_{ji}(\overline{Z})} \\ &= \theta_{ij}(Z) + \overline{\theta}_{ji}(Z), \end{aligned}$$

where $\overline{\theta}_{ij}$ means the complex-conjugate element of θ_{ij} in $(T^*M)^{\mathbb{C}}$. Thus,

$$\theta_{ij} + \overline{\theta}_{ji} = 0,$$

i.e., the connection matrix of D is skew-Hermitian. Conversely, if in a neighbourhood of any point of M there exists an orthonormal frame $e_1, ..., e_r$ for which the matrix of D is skew-Hermitian, then the connection D is metric with respect to the Hermitian metric g.

Let $E \to M$ be again a smooth complex vector bundle and h a Hermitian metric on E. Set $g = Re\, h$ and $\omega = Im\, h$. Then it follows from the identity $h(\xi, \eta) = \overline{h(\eta, \xi)}$ that $g(\xi, \eta) = g(\eta, \xi)$ and $\omega(\xi, \eta) = -\omega(\eta, \xi)$. Thus, ω is a skew-symmetric 2-form and g is an Euclidean metric on E considered as a real bundle; we will denote this real bundle by E^{re} for the moment. Let J be the complex structure on the fibres of E. Then the identity $h(\xi, \eta) = h(i\xi, i\eta)$ implies that $g(\xi, \eta) = g(J\xi, J\eta)$ and $\omega(\xi, \eta) = \omega(J\xi, J\eta)$. We also have $ig(\xi, \eta) - \omega(\xi, \eta) = ih(\xi, \eta) = h(i\xi, \eta) = g(J\xi, \eta) + i\omega(J\xi, \eta)$, hence it follows that $\omega(\xi, \eta) = -g(J\xi, \eta) = g(\xi, J\eta)$.

Obviously, if D is a connection on E that is metric with respect to h, then D is a connection on the real bundle E^{re} that is metric with respect to the Euclidean metric g. Moreover, as we have noted in Sect. 2.1, $DJs = JDs$ for every section s of E^{re}. Conversely, if D is a connection on E^{re} metric with respect to an Euclidean metric

g and if $DJs = JDs$, then D is a connection on the complex bundle E metric with respect to the Hermitian metric h on E defined by $h(\xi, \eta) = g(\xi, \eta) + ig(\xi, J\eta)$.

Problem 2.4.3 Prove the latter claim.

Denote the connection on the bundle $Hom(E^{re}, E^{re})$ induced by D with the same symbol. Then the identity $DJs = JDs$ means $DJ = 0$.

Apart from the identity $h(\xi, \eta) = g(\xi, \eta) + ig(\xi, J\eta)$, the relationship between the metrics h and g can be described as follows. Extend g to the complexification $E^{\mathbb{C}}$ of E by complex bilinearity. For the moment, denote the resulting $\mathbb{C}$-bilinear form by $g^{\mathbb{C}}$. Then $g^{\mathbb{C}}(\xi - iJ\xi, \eta + iJ\eta) = 2h(\xi, \eta)$. The vectors $\xi - iJ\xi$ and $\eta + iJ\eta$ in the left-hand side of this identity belong to the bundles $E^{1,0}$ and $E^{0,1}$, respectively. In order to interpret the right-hand side of the identity in terms of these bundles, recall that every Hermitian metric h_x, $x \in M$, can be considered as a $\mathbb{C}$-bilinear form $h_x : E_x \times \overline{E}_x \to \mathbb{C}$ ($\overline{E}_x$ is the real vector space E_x with the complex structure $-J_x$). In this way, h becomes a smooth section of the bundle $E^* \otimes_{\mathbb{C}} \overline{E}^*$. The complex bundles E and $\overline{E}$ are isomorphic to the bundles $E^{1,0}$ and $E^{0,1}$, respectively, via the maps $\xi \to \frac{1}{2}(\xi - iJ\xi)$ and $\xi \to \frac{1}{2}(\xi + iJ\xi)$. Therefore h_x can be considered as a $\mathbb{C}$-bilinear form on $E_x^{1,0} \times E_x^{0,1}$. Then the identity $g^{\mathbb{C}}(\xi - iJ\xi, \eta + iJ\eta) = 2h(\xi, \eta)$ means that transferring h onto $E^{1,0} \times E^{0,1}$, we have $h = 2g^{\mathbb{C}}|(E^{1,0} \times E^{0,1})$.

The form $g^{\mathbb{C}}$ on $E^{\mathbb{C}}$ possesses the following properties: (1) $g^{\mathbb{C}}$ is symmetric and $\mathbb{C}$-bilinear; (2) $g^{\mathbb{C}}(\overline{v}, \overline{w})) = \overline{g^{\mathbb{C}}(v, w)}$, $v, w \in E^{\mathbb{C}}$; (3) $g^{\mathbb{C}}(v, \overline{v}) > 0$ for every non-zero vector $v \in E^{\mathbb{C}}$; (4) $g^{\mathbb{C}}(v, w) = 0$ for $v, w \in E^{1,0}$ (and therefore a similar identity holds for $v, w \in E^{0,1}$). Conversely, every form on $E^{\mathbb{C}}$ with these properties is the complex bilinear extension of an Euclidean metric g on E such that $g(\xi, \eta) = g(J\xi, J\eta)$.

Let $E \to M$ and $F \to M$ be vector bundles and g^E, g^F metrics on these bundles. They yield in a natural way metrics on the bundles $E \oplus F$, $E \otimes F$, E^*, $Hom(E, F)$, $\Lambda^p E$, $\odot^p E$.

Recall that the induced metric $g^{E\otimes F}$ on $E \otimes F$ is uniquely determined by the identity $g^{E\otimes F}(e_1 \otimes f_1, e_2 \otimes f_2) = g^E(e_1, e_2) g^F(f_1, f_2)$.

The metric $g^{\otimes}$ on $\otimes^p E$ is defined via the identity

$$g^{\otimes}(a_1 \otimes ... \otimes a_p, b_1 \otimes ... \otimes b_p) = g^E(a_1, b_1) ... g^E(a_p, b_p).$$

The metrics $\Lambda^p E \subset \otimes^p E$ and $\odot^p E \subset \otimes^p E$ are multiples of the restrictions of $g^{\otimes}$; in particular, on $\Lambda^p E$, $g^{\Lambda}(a_1 \wedge ... \wedge a_p, b_1 \wedge ... \wedge b_p) = det[g^E(a_i, b_j]$.

Recall also that, since the form g_x^E is not degenerate for every $x \in M$, it determines an isomorphism f: $E_x \cong E_x^*$ given by $f(u)(v) = g_x^E(u, v)$ The metric on E_x can be transferred to E_x^* by means of this isomorphism, and in this way we get a metric g^{E^*} on the bundle E^* with the property that if $\alpha, \beta \in E_x^*$ and $e_1, ..., e_r$ is an orthonormal basis of E_x, then $g^*(\alpha, \beta) = \sum_{i=1}^{r} \alpha(e_i)\beta(e_i)$.

The induced metric on $Hom(E, F)$ is defined by means of the isomorphism $Hom(E, F) = E^* \otimes F$, where the metric on $E^* \otimes F$ is the one induced by g^{E^*}

and g^F. If $\varphi, \psi \in Hom(E, F)_x$ and $e_1, ..., e_r$ is an orthonormal basis of E_x, then $g^{Hom}(\varphi, \psi) = \sum_{i=1}^r g^F(\varphi(e_i), \psi(e_i))$.

Let $E \to N$ be a vector bundle, M a smooth manifold, and $f : M \to N$ a smooth map. Let $\widetilde{E} = f^*E \to M$ be the pull-back bundle of E, and $\widetilde{f} : \widetilde{E} \to E$ the canonical map $\widetilde{E} \ni (x, e) \to e$ where $x \in M, e \in E$ and $f(x) = \pi(e)$. The restriction of this map to every fibre is an isomorphism of vector spaces, and using it, we can transfer any metric on E to a metric on $\widetilde{E}$.

It is not hard to show that if D^E and D^F are connections on the bundles E and F compatible with the metrics g^E and g^F, then the connections induced by D^E and D^F on $E \oplus F, E \otimes F, E^*, Hom(E, F), \Lambda^p E, \odot^p E, \widetilde{E} = f^*E$ are compatible with the metrics defined above. For example, let $e_1, ..., e_r$ and $f_1' ..., f_r$ be orthonormal frames of E and F, respectively. Then $e_i \otimes f_j$, $1 \le i \le r$, $1 \le j \le s$, is an orthonormal frame of $E \otimes F$ and

$$\begin{aligned}
&g^{E\otimes F}(D_X^{E\otimes F}(e_i \otimes f_j), e_k \otimes f_l) + g^{E\otimes F}(e_i \otimes f_j, D_X^{E\otimes F}(e_k \otimes f_l)) \\
&= g^{E\otimes F}(D_X^E e_i \otimes f_j + e_i \otimes D_X^F f_j, e_k \otimes f_l) + g^{E\otimes F}(e_i \otimes f_j, D_X^E e_k \otimes f_l + e_k \otimes D_X^F f_l)) \\
&= [g^E(D_X^E e_i, e_k) + g^E(e_i, D_X^E e_k)]\delta_{jl} + \delta_{ik}[g^F(D_X^F f_j, f_l) + g^F(f_j, D_X^F f_l)] \\
&= 0, \quad X \in TM.
\end{aligned}$$

It follows that the connection $D^{E\otimes F}$ is compatible with the metric $g^{E\otimes F}$.

Proposition 2.4.2 *If $E \to M$ is a vector bundle over a paracompact manifold, then for any metric g on E there is a connection D compatible with g.*

Proof Let G be a metric on $\mathbb{K}^r$ and take a G-orthonormal basis $a = (a_1, ..., a_r)$ of $\mathbb{K}^r$. Let $e = (e_1, ..., e_r)$ be a g-orthonormal frame of E on an open set $U \subset M$. Denote by $\varphi_U : \pi^{-1}(U) \to U \times \mathbb{K}^r$ the trivialization of E determined by a and e. The map φ_U sends the vector $e_i(x)$, $x \in U$, to (x, a_i), $i = 1, ..., r$, hence φ_U is an isometry on the fibres. Now, transfer the trivial connection d on the trivial bundle $U \times \mathbb{K}^r$ to a connection on $\pi^{-1}(U)$ by means of the isometry φ_U, Since the connection d is compatible with the metric G, we get a connection D_U compatible with the metric g. Covering M by open sets U, and gluing together the connections D_U by means of a partition of unity, we get a connection compatible with g. □

Problem 2.4.4 Let a bundle E over a paracompact manifold be endowed with a metric. Show that if E admits a nowhere vanishing section ξ, E admits a metric connection D such that $D\xi = 0$.

Let D be a connection on a smooth manifold M. Set

$$T^D(X, Y) = D_X Y - D_Y X - [X, Y], \quad X, Y \in \chi(M).$$

It is not hard to see that $T^D(X, Y)$ is bilinear on the space of smooth functions on M. It follows that the value of the vector field $T^D(X, Y)$ at a point $p \in M$ depends only on the values X_p and Y_p of the vector fields X and Y at the point p. This allows one to define a tensor field T^D of type (2, 1) taking for $v, w \in T_pM$ vector fields X, Y with $X_p = v$, $Y_p = w$ and setting $T^D(v, w) = T^D(X, Y)_p$. It is clear that the tensor T^D is skew-symmetric, hence it is a section of the bundle $\Lambda^2 T^*M \otimes TM$. This tensor is called the **torsion** of the connection D.

Proposition 2.4.3 *On every Riemannian manifold (M, g), there exists a unique metric connection ∇ with vanishing torsion. This connection is determined by the formula*

$$g(\nabla_X Y, Z) = \frac{1}{2}[Xg(Y, Z) + Yg(Z, X) - Zg(X, Y) \\ +g(Z, [X, Y]) + g(Y, [Z, X]) - g(X, [Y, Z]), \quad X, Y, Z \in \chi(M). \tag{2.6}$$

A proof of this proposition can be found in any standard textbook on Riemannian geometry, see, for example, O'Neill [21], Chap. 3, Sect. 2, Theorem 11 or Kobayashi and Nomizu [14], vol. I, Chap. IV, Sect. 2, Theorem 2.2. Here, we only give a **sketch of a proof of existence**.

Denote the right-hand side of (2.6) by $F(X, Y, Z)$. For fixed vector fields X and Y, the function $Z \to F(X, Y, Z)$ is linear on the space of smooth functions on M. Therefore the values of this function at a point $p \in M$ depend only on the value Z_p of Z at p. It follows that there exists a unique vector field, which we denote by $\nabla_X Y$, such that $2g(\nabla_X Y, Z) = F(X, Y, Z)$. The map $X \to \nabla_X Y$ is linear on the space of smooth functions, hence its value at $p \in M$ depends only on the value X_p of X at p. Now, using (2.6), one can check that $\nabla_X Y$ is a connection which is metric and has vanishing torsion.

Definition 2.4.3 The connection defined in Proposition 2.4.3 is called the Levi-Civita connection of the Riemannian manifold (M, g). Identity (2.6) is called the Koszul formula.

As we have noted, the torsion of any connection is a skew-symmetric tensor of type (2, 1). If T is an arbitrary skew-symmetric tensor of type (2, 1) on M, then there exists a unique connection D on M which is metric and whose torsion is equal to T. This connection is determined by the identity

$$g(D_X Y, Z) = g(\nabla_X Y, Z) + \frac{1}{2}[g(Z, T(X, Y)) + g(Y, T(Z, X)) - g(X, T(Y, Z))],$$

where ∇ is the Levi-Civita connection.

Note that we can construct infinitely many skew-symmetric tensors of type (2, 1) on M. For example, define such tensors on a coordinate neighbourhood of M, then extend these locally defined tensors to the whole manifold M. Therefore every Riemannian manifold admits infinitely many metric connections.

2.5 The Canonical Connection on a Hermitian Holomorphic Vector Bundle (Chern's Connection)

As we have seen, on a smooth vector bundle there are many connections even if we want them to be compatible with a given metric. A natural choice of a connection does not exist. But in the case when the bundle is holomorphic and endowed with a Hermitian metric, two natural conditions can be imposed that lead to a canonical choice of a connection on the bundle.

Let $\pi : E \to M$ be a holomorphic vector bundle. Denote the complex structure of the complex manifold by $J : TM \to TM$. For $v \in TM$, let $v^{1,0}$ and $v^{0,1}$ be the projections of v to $T^{1,0}M$ and $T^{0,1}M$: $v^{1,0} = \frac{1}{2}(v - iJv)$, $v^{0,1} = \frac{1}{2}(v + iJv)$. If D is a connection on E, its $(1, 0)$ and $(0, 1)$ parts D' and D'' are defined by

$$D'_v s = D_{v^{1,0}} s, \quad D''_v s = D_{v^{0,1}} s, \quad s \in A^0(E).$$

Let $A^{p,q}(E)$ be the space of sections of the bundle $\Lambda^{p,q}T^*M \otimes E$, i.e., the space of (p, q)- forms on M with values in E. Then D' and D'' map $A^0(E)$ into $A^{1,0}(E)$ and $A^{0,1}(E)$, respectively:

$$D' : A^0(E) \to A^{1,0}(E), \quad D'' : A^0(E) \to A^{0,1}(E).$$

These maps are $\mathbb{C}$-linear and satisfy the Leibniz rule of the form

$$D'(fs) = \partial f \otimes s + fD's, \quad D''(fs) = \overline{\partial} f \otimes s + fD''s, \quad f \text{ a smooth function } M.$$

Recall that ∂f and $\overline{\partial} f$ are 1-forms of type $(1, 0)$ and $(0, 1)$ defined by $\partial f(v) = df(v^{1,0}) = v^{1,0}(f)$, $\overline{\partial} f(v) = df(v^{0,1}) = v^{0,1}(f)$.

Also, since E is a holomorphic bundle, the $\overline{\partial}$ operator is well-defined on the sections of E: $\overline{\partial} = \overline{\partial}_E : A^0 \to A^{0,1}(E)$.

Definition 2.5.1 A connection D on a holomorphic vector bundle E is called consistent or compatible with its complex structure if $D'' = \overline{\partial}$.

The condition $D'' = \overline{\partial}$ can be stated as follows: A smooth (local) section s of E is holomorphic if and only if $D_w s = 0$ for every $(0, 1)$-tangent vector w of M. For, suppose that the latter condition holds. Let $s_1, ..., s_r$ be a holomorphic frame of E. Then for any smooth section s of E, we have $s = \sum_{i=1}^r \lambda_i s_i$ where λ_i are smooth functions, and $D_w s = \sum_i w(\lambda_i)s_i + \sum_i \lambda_i D_w s_i = \sum_i w(\lambda_i)s_i = \overline{\partial} s(w)$. Therefore $D'' = \overline{\partial}$. The converse claim is obvious (recall that a smooth section s is holomorphic exactly when $\overline{\partial} s = 0$).

Note that the condition for consistency of a connection D with the complex structure of a holomorphic bundle E means that the connection matrix θ with respect to any holomorphic frame $e_1, ..., e_r$ is of type $(1, 0)$. Indeed, let $\sigma = \sum_{i=1}^{r} \sigma_i e_i$ be a smooth section of E. Then

$$D\sigma = \sum_{j=1}^{r}(d\sigma_j + \sum_{i=1}^{r} \sigma_i \theta_{ij}) \otimes e_j = \sum_j (\partial\sigma_j + \sum_i \sigma_i \theta_{ij}^{1,0}) \otimes e_j + \sum_j (\overline{\partial}\sigma_j + \sum_i \sigma_i \theta_{ij}^{0,1}) \otimes e_j,$$

where $\theta_{ij}^{1,0}$ and $\theta_{ij}^{0,1}$ are the $(1, 0)$ and $(0, 1)$ parts of the connection forms θ_{ij}. Hence

$$D'\sigma = \sum_j (\partial\sigma_j + \sum_i \sigma_i \theta_{ij}^{1,0}) \otimes e_j, \quad D''\sigma = \sum_j (\overline{\partial}\sigma_j + \sum_i \sigma_i \theta_{ij}^{0,1}) \otimes e_j.$$

Furthermore, $\overline{\partial}\sigma = \sum_j \overline{\partial}\sigma_j \otimes e_j$. Therefore if $D'' = \overline{\partial}$, then $\theta_{ij}^{0,1} = 0$, hence $\theta_{ij} = \theta_{ij}^{1,0}$ is of type $(1, 0)$. Conversely, if the forms θ_{ij} are of type $(1, 0)$, then $D''\sigma = \sum_j \overline{\partial}\sigma_j \otimes e_j = \overline{\partial}\sigma$, i.e., the connection D is consistent with the complex structure of E.

Proposition 2.5.1 *Let $\pi : E \to M$ be a holomorphic vector bundle endowed with a Hermitian metric h. Then there exists a unique metric connection D consistent with the complex structure of E.*

Proof Let $e_1, ..., e_r$ be a holomorphic frame of E. Set $h_{ij} = h(e_i, e_j)$, and denote the matrix $[h_{ij}]$ by h.

Suppose that D is a connection on E that has the properties stated in the proposition. Let $\theta = [\theta_{ij}]$ be the matrix of D with respect to the holomorphic frame $e_1, ..., e_r$. Since D is consistent with h,

$$dh_{ij} = dh(e_i, e_j) = h(De_i, e_j) + h(e_i, De_j) = \sum_{k=1}^{r}(\theta_{ik} h_{kj} + h_{ik} \overline{\theta}_{jk}),$$

or, in matrix form, $dh = \theta h + h\, {}^t\overline{\theta}$. Therefore

$$\partial h = \theta^{1,0} h + h\, {}^t\overline{\theta}^{1,0}, \quad \overline{\partial} h = \theta^{0,1} h + h\, {}^t\overline{\theta}^{0,1}.$$

Note that the identity $\overline{\theta}^{1,0} + \overline{\theta}^{0,1} = \overline{\theta} = \overline{\theta^{1,0}} + \overline{\theta^{0,1}}$ implies $\overline{\theta}^{1,0} = \overline{\theta^{0,1}}$ and $\overline{\theta}^{0,1} = \overline{\theta^{1,0}}$. Since D is consistent with the complex structure of E, $\theta^{0,1} = 0$, hence $\partial h = \theta^{1,0} h$ and we get

$$\theta = \theta^{1,0} = \partial h \cdot h^{-1}.$$

This formula proves the uniqueness of D. To prove the existence of a connection with the desired properties, we check that for the matrices locally defined by $\theta = \partial h \cdot h^{-1}$ the rule (2.2) for changing the connection matrices under changing the frames holds. Let $e'_1, ..., e'_r$ be a holomorphic frame of E and $e'_i = \sum_{j=1}^{r} g_{ij} e_j$; the coefficients g_{ij} are holomorphic functions. Set $h'_{ij} = h(e'_i, e'_j)$, $h' = [h'_{ij}]$, $g = [g_{ij}]$. Then $h' =$

$gh\,{}^t\overline{g}$. Since the functions g_{ij} are holomorphic, $dg_{ij} = \partial g_{ij}$ and $\partial \overline{g}_{ij} = \overline{\overline{\partial} g_{ij}} = 0$. Hence

$$\theta' = \partial h' \cdot (h')^{-1} = (dg \cdot h \cdot {}^t\overline{g} + g \cdot \partial h \cdot {}^t\overline{g}) \cdot {}^t\overline{g}^{-1} \cdot h^{-1} \cdot g^{-1} = dg \cdot g^{-1} + g \cdot \theta \cdot g^{-1}.$$

This shows that there exists a connection D on E whose matrix with respect to a holomorphic frame $e_1, ..., e_r$ is given by $\theta = \partial h \cdot h^{-1}$, $h = [h(e_i, e_j)]$. Since the connection matrix θ is of type $(1, 0)$, the connection D is consistent with the complex structure of E. Furthermore,

$$h(De_i, e_j) + h(e_i, De_j) = \sum_{k=1}^{r} (\theta_{ik} h_{kj} + \overline{\theta}_{jk} \overline{h}_{ki})$$
$$= \partial h_{ij} + \overline{\partial h_{ji}} = \partial h_{ij} + \overline{\partial}\, \overline{h}_{ji} = \partial h_{ij} + \overline{\partial} h_{ij} = dh_{ij}.$$

It easily follows that the connection D is consistent with the metric h. □

Definition 2.5.2 If E is a Hermitian holomorphic vector bundle, the unique connection consistent with the metric and the complex structure of E is called the canonical connection on E or the Chern connection.

Proposition 2.5.2 *Let $E \to M$ be a Hermitian holomorphic vector bundle and $F \subset E$ a holomorphic subbundle. Let $p_F : E \to F$ be the projection with respect to the orthogonal decomposition $E = F \oplus F^{\perp}$. Then if D is the canonical connection of E, $p_F \circ D|A^0(F)$ is the canonical connection of the holomorphic bundle F endowed with the Hermitian metric $h|F$.*

Proof It is easy to check that $D^F = p_F \circ D|A^0(F)$ is a connection on the bundle F. Note also that the restriction of the operator $\overline{\partial}_E$ to the sections of F coincides with the operator $\overline{\partial}_F$. Indeed, for every point p of M, there is a holomorphic frame $e_1, ..., e_r$ of E on a neighbourhood of p such that $e_1, ..., e_s$ is a frame of F. Every smooth section ξ of F locally has the form $\xi = \sum_{i=1}^{s} \xi_i \otimes e_i$, hence $\overline{\partial}_F = \sum_{i=1}^{s} \overline{\partial} \xi_i \otimes e_i = \overline{\partial}_E \xi$. Then $(D^F)'' \xi = p_F(D'' \xi) = p_F(\overline{\partial}_E \xi) = p_F(\overline{\partial}_F \xi) = \overline{\partial}_F \xi$, i.e., the connection D^F is consistent with the complex structure of F. Furthermore, for every two smooth sections ξ and η of F, $dh(\xi, \eta) = h(D\xi, \eta) + h(\xi, D\eta) = h(p_F(D\xi), \eta) + h(\xi, p_F(D\eta))$, i.e., the connection D^F is consistent with the metric of the bundle F. □

Proposition 2.5.3 *Let $E \to M$ and $F \to M$ be Hermitian holomorphic vector bundles with metrics h^E and h^F, and canonical connections D^E and D^F. Endow the bundles $E \oplus F$, $E \otimes F$, E^*, $Hom(E, F)$, $\Lambda^p E$, $\odot^p E$ with the Hermitian metrics induced by h^E and h^F. Then the connection on each of these bundles induced by D^E and D^F coincides with the canonical connection of the bundle.*

Proof We prove the proposition only for the bundle $E \otimes F$. The proof for the other bundles is similar.

Let $e_1, ..., e_r$ and $f_1, ..., f_s$ be holomorphic frames of E and F. Then $e_i \otimes f_j$, $i = 1, ..., r$, $j = 1, ..., s$, is a holomorphic frame of the bundle $E \otimes F$. Hence if s is a smooth section of this bundle, then $s = \sum_{i,j} \lambda_{ij} e_i \otimes f_j$ and

$$(D^{E\otimes F})'' s = \sum_{i,j} [\overline{\partial}\lambda_{ij} \otimes e_i \otimes f_j + \lambda_{ij}(D^E)'' e_i \otimes f_j + \lambda_{ij} e_i \otimes (D^F)'' f_j]$$
$$= \sum_{i,j} \overline{\partial}\lambda_{ij} \otimes e_i \otimes f_j = \overline{\partial}s.$$

Therefore the connection $D^{E\otimes F}$ is consistent with the complex structure of $E \otimes F$. Moreover, as we have mentioned in the preceding paragraph, this connection is consistent with the metric $h^{E\otimes F}$ yielded by h^E and h^F. □

Proposition 2.5.4 *Let $E \to N$ be a Hermitian holomorphic vector bundle, M a complex manifold, and $f : M \to N$ a holomorphic map. Endow the pull-back bundle $\widetilde{E} = f^*E$ with the Hermitian metric induced by the metric of E. Then the canonical connection of $\widetilde{E}$ coincides with the connection induced by the canonical connection of E.*

The proof is left to the reader.

Now, let M be a complex manifold and D a connection on M, i.e., on the complex bundle $T^h M$. As we have noted in Chap. 1, Sect. 1.2, if we consider M as a smooth manifold its tangent bundle TM is canonically isomorphic to the bundle $T^h M$ considered as a smooth real vector bundle. Hence D can be considered as a connection on the real bundle TM such that if $J : TM \to TM$ is the complex structure of M, $DJX = JDX$ for every smooth vector field X on M (see Sect. 2.1).

The connection D can be extended by complex linearity to a connection on the complex bundle $T^{\mathbb{C}} M$. Suppose D is consistent with the complex structure of M, i.e., with the complex structure of the holomorphic bundle $T^h M$. Let $z_1, ..., z_n$ be complex coordinates of M. For convenience, set $Z_k = \dfrac{\partial}{\partial z_k}$, $k = 1, ..., n$. The vector fields Z_k are holomorphic, hence

$$D_{\overline{Z}_l} Z_k = 0 \text{ and } D_{Z_k} \overline{Z}_l = \overline{D_{\overline{Z}_k} Z_l} = 0, \quad k, l = 1, ..., n.$$

Then, for the torsion T of D, we have $T(Z_k, \overline{Z}_l) = 0$. Since $\{Z_k\}$ and $\{\overline{Z}_l\}$ are frames of $T^{1,0}M$ and $T^{0,1}M$, it follows that $T(X - iJX, Y + iJY) = 0$ for every $X, Y \in TM$ (i.e., the $(1, 1)$-part of T vanishes). The latter identity is equivalent to the identity $T(X, Y) + T(JX, JY) = 0$. Conversely, if the latter identity holds, the connection D is consistent with the complex structure of $T^h M$. To prove this, note first that the identity $DJX = JDX$ implies that for every $Y \in TM$

$$D_{Y+iJY}(X - iJX) = (D_Y X - iJ D_Y X) + i(D_{JY} X - iJ D_{JY} X) \in T^{1,0}M$$

$$D_{Y-iJY}(X + iJX) = (D_Y X + iJ D_Y X) - i(D_{JY} X + iJ D_{JY} X) \in T^{0,1}M.$$

In particular, the vector field $D_{\overline{Z}_k} Z_l$ is of type $(1, 0)$ and $D_{Z_l} \overline{Z}_k$ is of type $(0, 1)$. Moreover, according to our assumption, $D_{\overline{Z}_k} Z_l - D_{Z_l} \overline{Z}_k = T(\overline{Z}_k, Z_l) = 0$, hence $D_{\overline{Z}_k} Z_l = D_{Z_l} \overline{Z}_k$ is of type $(0, 1)$. Therefore $D_{\overline{Z}_k} Z_l = 0$. Thus $D_W Z_l = 0$ for $W \in T^{0,1}M$, $l = 1, ..., n$. Every smooth section Z of $T^h M$ is of the form $Z = \sum_{l=1}^n \lambda_l Z_l$ where λ_l are smooth functions. Then $D_W Z = \sum_{l=1}^n W(\lambda_l) Z_l = \sum_{l=1}^n (\overline{\partial}\lambda_l)(W) Z_l = (\overline{\partial} Z)(W)$. Hence $D'' = \overline{\partial}$, i.e., the connection D is consistent with the complex structure of $T^h M$.

The above reasoning gives the following result.

Proposition 2.5.5 *A connection D on a complex manifold M (i.e., on $T^h M$) is consistent with the complex structure J of M if and only if the following two conditions are satisfied:*

(1) $DJ = 0$ where $J : TM \to TM$ is the complex structure of M

(2) The torsion T of D obeys the identity

$$T(X, Y) + T(JX, JY) = 0, \quad X, Y \in TM.$$

Note that the identity $DJ = 0$ implies

$$D_{Y-iJY}(X - iJX) = (D_Y X - iJ D_Y X) - i(D_{JY} X - iJ D_{JY} X) \in T^{1,0}M.$$

Hence, if V and W are $(1, 0)$-vector fields on M, the vector field $D_V W$ is also of type $(1, 0)$. The vector fields V and W locally have the representations $V = \sum_{k=1}^n \lambda_k Z_k$ and $W = \sum_{l=1}^n \mu_l Z_l$, where as above $Z_k = \dfrac{\partial}{\partial z_k}$, $k = 1, ..., n$. Then $[V, W] = \sum_{k,l=1}^n (\lambda_k Z_k(\mu_l) Z_l - \mu_l Z_l(\lambda_k) Z_k)$, hence the vector filed $[V, W]$ is of type $(1, 0)$. It follows that the vector field $T(V, W) = D_V W - D_W V - [V, W]$ is of type $(1, 0)$.

The torsion T of the connection D on $T^{\mathbb{C}}M$ is the $\mathbb{C}$-bilinear extension of the torsion $T : TM \times TM \to TM$ of the connection on TM. Hence $T(\overline{V}, \overline{W}) = \overline{T(V, W)}$. If T satisfies the condition (2) in Proposition 2.5.5, then $T(V, \overline{W}) = T(\overline{V}, W) = 0$. Therefore in this case the torsion of D is uniquely determined by its values $T(V, W)$ for $V, W \in T^{1,0}M$.

Now, let the complex manifold M be endowed with a Hermitian metric h. As we have noted in Sect. 2.4, the requirement that D be a connection on $T^h M$ consistent with the metric h is equivalent to the requirement that D be a connection on TM consistent with the metric $g = Reh$ and such that $DJ = 0$. This and Proposition 2.5.5 imply the following.

Proposition 2.5.6 *The canonical connection of a Hermitian complex manifold M is the unique connection D on TM such that:*

(1) $DJ = 0$ where $J : TM \to TM$ is the complex structure of M

(2) The torsion T of D satisfies the identity

$$T(X, Y) + T(JX, JY) = 0, \quad X, Y \in TM$$

(3) D is consistent with the Riemannian metric g on M that is the real part of the Hermitian metric.

Let M be an almost complex manifold with almost complex structure J. If M is endowed with a Riemannian metric g compatible with J in the sense that $g(JX, JY) = g(X, Y)$, $X, Y \in TM$, then M is said to be an almost Hermitian manifold. Every Hermitian complex manifold with metric h is almost Hermitian with respect to the complex structure of M and the Riemannian metric $g = Re\, h$. For more information on almost Hermitian manifolds and examples of such manifolds that are not complex, see Kobayashi and Nomizu [14], vol. II, Chap. IX.

It has been observed by Lichnerowicz that the three conditions of Proposition 2.5.6 define a unique connection on every almost Hermitian manifold.

Proposition 2.5.7 *On every almost Hermitian manifold (M, g, J), there exists a unique connection D such that:*

(1) $DJ = 0$

(2) The torsion T of D satisfies the identity

$$T(X, Y) + T(JX, JY) = 0, \quad X, Y \in TM$$

(3) D is consistent with the metric g.

This connection is given by the identity

$$\begin{aligned} g(D_X Y, Z) &= g(\nabla_X Y, Z) + \tfrac{1}{2} g((\nabla_X J)(Y), JZ) \\ &\quad + \tfrac{1}{4} g(-(\nabla_Y J)(Z) - (\nabla_{JY} J)(JZ) + (\nabla_Z J)(Y) + (\nabla_{JZ} J)(JY), JX), \\ &\qquad X, Y, Z \in \chi(M), \end{aligned}$$

where ∇ is the Levi-Civita connection of the metric g.

A proof of this proposition can be seen in the paper by Gray et al. [8] cited in the list of references.

It follows immediately from the formula giving the relationship between the Chern and the Levi-Civita connections, that if $\nabla J = 0$, the two connections coincide. The converse is a direct consequence of (1) of Proposition 2.5.6. Suppose that $\nabla J = 0$.

Then we have for the Nijenhuis tensor

$$\begin{aligned} N(X,Y) &= -[X,Y] + [JX,JY] - J[JX,Y] - J[X,JY] \\ &= -J(\nabla_X J)(Y) + J(\nabla_Y J)(X) + (\nabla_{JX} J)(Y) - (\nabla_{JY} J)(X) = 0. \end{aligned}$$

Hence the almost complex structure J is integrable, i.e., M is a complex manifold. Thus, if $\nabla J = 0$, then (M, g, J) is a Hermitian complex manifold. A Hermitian complex manifold for which $\nabla J = 0$ is called **a Kähler manifold**. It follows from Propositions 2.5.6 and 2.4.3 that the Kähler condition is equivalent to the condition that the torsion of the Chern connection vanishes.

Another condition that is easier to be checked is the following. Let M be a Hermitian complex manifold. Then $\omega = Im\, h$ is a skew-symmetric 2-form such that $\omega(X,Y) = g(X, JY)$, $X, Y \in TM$, where $g = Re\, h$ (see Sect. 2.4). This form is called **the fundamental or the Kähler form** of M. It is an important result in complex geometry that M is Kähler if and only if its fundamental form is closed: $d\omega = 0$, see Moroianu [18], Part 2, Sect. 5, Theorem 5.5 or Kobayashi and Nomizu [14], vol. II, Chap. IX, Sect. 4, Theorem 4.3. Therefore every Riemann surface (one-dimensional complex manifold) endowed with a Hermitian metric is Kähler.

2.6 Exterior Differential Operator on Forms with Values in a Vector Bundle Determined by a Connection on the Bundle

Let $\pi : E \to M$ be a smooth $\mathbb{K}$-vector bundle. Denote by $A^p(\mathbb{K}) = A^p(M, \mathbb{K})$ the space of $\mathbb{K}$-valued p-forms on the manifold M.

Let $\varphi \in A^q(E)$ be a q-form with values in E. If $e_1, ..., e_r$ is a frame of E, then φ can be uniquely represented as $\varphi = \sum_{i=1}^r \varphi_i \otimes e_i$ where $\varphi_i \in A^q(\mathbb{K})$. Let $\omega \in A^p(\mathbb{K})$. Define a form $\omega \wedge \varphi \in A^{p+q}(E)$ by means of the identity

$$\omega \wedge \varphi = \sum_{i=1}^{r} (\omega \wedge \varphi_i) \otimes e_i. \tag{2.7}$$

It can be easily verified that this form does not depend on the choice of the frame $e_1, ..., e_r$. Clearly, if $p = 0$, i.e., ω is a smooth function, then $\omega \wedge \varphi = \omega\varphi \in A^q(E)$; if $q = 0$, i.e., φ is a section of E, $\omega \wedge \varphi = \omega \otimes \varphi \in \Lambda^p T^*M \otimes E = A^p(E)$. For $p \geq 1, q \geq 1$, and $X_1, ..., X_{p+q} \in T_xM$, we have

$$(\omega\wedge\varphi)(X_1, ..., X_{p+q}) = \frac{1}{p!q!} \sum_{\sigma} \varepsilon_\sigma \omega(X_{\sigma(1)}, ..., X_{\sigma(p)}) \varphi(X_{\sigma(p+1)}, ..., X_{\sigma(p+q)}),$$

where the summation is over all permutations σ of the numbers $\{1, ..., p+q\}$ and ε_σ is the signature of σ.

Set also

$$\varphi \wedge \omega = \sum_{i=1}^{r} (\varphi_i \wedge \omega) \otimes e_i = (-1)^{pq} \omega \wedge \varphi.$$

Problem 2.6.1 Prove that if $\omega \in A^p(\mathbb{K})$ and $\varphi \in A^1(E)$, then for $X_1, \ldots, X_{p+1} \in T_x M$

$$(\omega \wedge \varphi)(X_1, \ldots, X_{p+1}) = (-1)^p \sum_{i=1}^{p+1} (-1)^{i+1} \omega(X_1, \ldots, \widehat{X_i}, \ldots, X_{p+1}) \varphi(X_i).$$

Recall that the trivial connection d on the trivial bundle over a manifold M, differentiation of functions, gives rise to a linear operator sending the scalar-valued p-forms on M into $(p+1)$-forms and such that

$$d(\varphi \cdot f) = d\varphi \cdot f + (-1)^p \varphi \wedge df, \tag{2.8}$$

for every p-form φ and every smooth function f.

By analogy, in the case of an arbitrary vector bundle $\pi : E \to M$, we can consider the problem of defining an operator $A^p(E) \to A^{p+1}(E)$ with similar properties by means of an arbitrary connection D on the bundle E.

Let $\varphi = \sum_{i=1}^{r} \varphi_i \otimes e_i$ be the representation of a form $\varphi \in A^p(E)$ in terms of a frame $e_1, \ldots, e_r$ of E, where φ_i are $\mathbb{K}$-valued p-forms. By analogy with formula (2.8), where f is in fact a section of the trivial bundle, we set

$$\begin{aligned} D\varphi &= \sum_{i=1}^{r} [d\varphi_i \otimes e_i + (-1)^p \varphi_i \wedge De_i] \\ &= \sum_{j=1}^{r} [d\varphi_j + (-1)^p \sum_{i=1}^{r} \varphi_i \wedge \theta_{ij}] \otimes e_j, \end{aligned} \tag{2.9}$$

where $[\theta_{ij}]$ is the matrix of the connection D with respect to the frame $e_1, \ldots, e_r$. It is not hard to verify that the right-hand side of this identity does not depend on the choice of the frame $e_1, \ldots, e_r$. Therefore formula (2.9) defines an operator

$$D : A^p(E) \to A^{p+1}(E),$$

which is $\mathbb{K}$-linear and such that for every $\mathbb{K}$-valued p-form ω on M and every section σ of E

$$D(\omega \otimes \sigma) = d\omega \otimes \sigma + (-1)^p \omega \wedge D\sigma.$$

These properties uniquely determine the operator D since every form in $A^p(E)$ is the sum of forms of type $\omega \otimes \sigma$.

Clearly, if E is the trivial bundle and D is the trivial connection d, the operator defined above coincides with the operator of exterior differentiation on scalar-valued forms. Because of the analogy with the operator of exterior differentiation, the operator $D : A^p(E) \to A^{p+1}(E)$ is often denoted by d^D or just by d when the connection

D is understood. It is said that $D\varphi$, $\varphi \in A^p(E)$, is the differential of φ with respect to the connection D.

Note that (2.8) and (2.9) imply that if $\omega \in A^p(\mathbb{K})$ and $\varphi \in A^q(E)$

$$D(\omega \wedge \varphi) = d\omega \wedge \varphi + (-1)^p \omega \wedge D\varphi.$$

Problem 2.6.2 Prove that if $X_1, ..., X_{p+1}$ are vector fields on M and $\varphi \in A^p(E)$, then

$$(D\varphi)(X_1, ..., X_{p+1}) = \sum_{k=1}^{p+1} (-1)^{k+1} D_{X_k}(\varphi(X_1, ..., \widehat{X}_k, ..., X_{p+1})) \\ + \sum_{k<l} (-1)^{k+l} \varphi([X_k, X_l], X_1, ..., \widehat{X}_k, ..., \widehat{X}_l, ..., X_{p+1}).$$

This formula is similar to the well-known formula for the exterior derivative of a scalar-valued form, see the next paragraph.

2.7 Curvature of a Connection

A very important property of the exterior differentiation operator d is that $d^2 = 0$ (the theory of de Rham cohomology is based on it, for example). Therefore we now compute the square of the operator $D : A^p(E) \to A^{p+1}(E)$.

By Leibniz's rule, for every smooth section σ of E on an open subset of M and for every smooth function f on it

$$D^2(f\sigma) = D(df \otimes \sigma + f D\sigma) = d^2 f \otimes \sigma - df \wedge D\sigma + df \wedge D\sigma + f D(D\sigma) = f D^2 \sigma.$$

Hence if two sections s and σ coincide at a point $x \in M$, then $(D^2 s)(x) = (D^2 \sigma)(x)$. Indeed, let $e_1, ..., e_r$ be a frame of E in a neighbourhood of the point x and let $s = \sum_{i=1}^r s_i e_i$, $\sigma = \sum_{i=1}^r \sigma_i e_i$. The coefficients $s_i(x) = \sigma_i(x)$ since $s(x) = \sigma(x)$. Then

$$(D^2 s)(x) = (D^2(\sum_{i=1}^r s_i e_i))(x) = \sum_{i=1}^r s_i(x)(D^2 e_i)(x) = \sum_{i=1}^r \sigma_i(x)(D^2 e_i)(x) = (D^2 \sigma)(x).$$

This observation allows one to define a global smooth section of the bundle $\Lambda^2 T^* M \otimes Hom(E, E)$, i.e., a 2-form with values in the bundle $Hom(E, E)$, as follows. Let $v, w \in T_x M$ and $e \in E_x$. Take any section σ of E such that $\sigma(x) = e$ and set

$$R(v, w)e = (D^2 \sigma)_x(v, w).$$

Definition 2.7.1 The section R of $\Lambda^2 T^* M \otimes Hom(E, E)$ defined in this way is called the curvature of the connection D or the curvature tensor.

For $v, w \in TM^{\mathbb{C}}$ and $e \in E^{\mathbb{C}}$, define $R(v, w)e$ by complex trilinear extension. Clearly, if E is a complex bundle, $R(v, w)e \in E$ for $v, w \in TM^{\mathbb{C}}$ and $e \in E$.

Problem 2.7.1 Let $e_1, ..., e_r$ be a frame of E in a neighbourhood of a point $x \in M$. Prove that if $\varphi = \sum_{i=1}^{r} \varphi_i \otimes e_i \in A^p(E)$ and $X_1, ..., X_{p+2} \in T_x M$, then

$$(D^2\varphi)(X_1, ..., X_{p+2}) = \sum_{i=1}^{r} \frac{1}{p!2!} \sum_{\sigma} \varepsilon_\sigma \varphi_i(X_{\sigma(1)}, ..., X_{\sigma(p)}) R(X_{\sigma(p+1)}, X_{\sigma(p+2)}) e_i$$
$$= \frac{1}{p!2!} \sum_{\sigma} \varepsilon_\sigma R(X_{\sigma(p+1)}, X_{\sigma(p+2)})(\varphi(X_{\sigma(1)}, ..., X_{\sigma(p)})),$$

where summation is taken over all permutations σ of $\{1, ..., p+2\}$ and ε_σ is the signature of σ.

If $e = (e_1, ..., e_r)$ is a frame of E, we have

$$D^2 e_i = \sum_{j=1}^{r} \Theta_{ij} \otimes e_j,$$

where Θ_{ij} are $\mathbb{K}$-valued 2-forms. The matrix $\Theta_e = [\Theta_{ij}]$ is called the **curvature matrix** of D with respect to the frame e. If $e' = (e'_1, ..., e'_r)$ is another frame and $e'_i = \sum_{j=1}^{r} g_{ij} e_j$, then

$$\sum_{j=1}^{r} \Theta'_{ij} \otimes e'_j = D^2 e'_i = \sum_{j,k} g_{ij} \Theta_{jk} \otimes e_k.$$

Hence

$$\Theta_{e'} = g \cdot \Theta_e \cdot g^{-1}, \quad g = [g_{ij}].$$

If E is a line bundle, then $g \in \mathbb{K} \setminus \{0\}$, therefore $\Theta_{e'} = \Theta_e$. In this case, the form Θ locally defined as Θ_e is, in fact, globally defined on M.

The sections L_{ij}, $i, j = 1, ..., r$, of the bundle $Hom(E, E)$ defined by $L_{ij}(e_k) = \delta_{ik} e_j$, $k = 1, ..., r$, constitute a frame of $Hom(E, E)$. If $v, w \in TM$, $R(v, w)e_k = (D^2 e_k)(v, w) = \sum_{j=1}^{r} \Theta_{kj}(v, w) e_j$. Hence

$$R = \sum_{i,j=1}^{r} \Theta_{ij} \otimes L_{ij}.$$

The curvature matrix is simply expressed in terms of the matrix of the connection. Indeed,

$$\begin{aligned} D^2 e_i = D(De_i) = D(\sum_{j=1}^{r} \theta_{ij} \otimes e_j) &= \sum_{j=1}^{r} [d\theta_{ij} \otimes e_j - \theta_{ij} \wedge De_j] \\ &= \sum_{j=1}^{r} d\theta_{ij} \otimes e_j - \sum_{j,k=1}^{r} \theta_{ij} \wedge \theta_{jk} \otimes e_k \\ &= \sum_{k=1}^{r} [d\theta_{ik} - \sum_{j=1}^{r} \theta_{ij} \wedge \theta_{jk}] \otimes e_k. \end{aligned}$$

Therefore

$$\Theta_{ik} = d\theta_{ik} - \sum_{j=1}^{r} \theta_{ij} \wedge \theta_{jk},$$

or in matrix notation

$$\Theta_e = d\theta_e - \theta_e \wedge \theta_e.$$

This identity is called **Cartan's structure equation.**

Suppose that D is a metric connection. Then the connection matrix θ_{ij} with respect to an orthonormal frame is skew-symmetric as remarked in Sect. 2.4. It follows from the Cartan structure equation that the curvature matrix of D is also skew-symmetric. Similarly, if the metric is Hermitian, the curvature matrix is skew-Hermitian.

Problem 2.7.2 Let D be a connection on a vector bundle E and D^* the induced connection on the dual bundle E^*. Prove that if Θ is the curvature matrix of D with respect to a frame of E, $\Theta^* = -\Theta^t$ is the curvature matrix of D^* with respect to the dual frame.

Problem 2.7.3 Let $E \to N$ be a vector bundle with a connection D and $f : M \to N$ a smooth map. Denote the induced connection on the pull-back bundle f^*E by $\widetilde{D}$. Prove that if Θ is the curvature matrix of D with respect to a frame $e_1, ..., e_r$ of E, $\widetilde{\Theta} = f^*\Theta$ is the curvature matrix of $\widetilde{D}$ with respect to the frame $(\widetilde{e}_1 = e_1 \circ f, ..., \widetilde{e}_r = e_r \circ f)$ of the bundle f^*E.

The matrix of the trivial connection on the trivial bundle is zero, and Cartan's structure equation implies that the curvature of this connection is zero. Connections with zero curvature are called **flat**. We shall later show that a vector bundle possessing a connection D of zero curvature is locally isomorphic to the trivial bundle by an isomorphism sending the connection D to the trivial connection. Note that if $De_i = 0$ for every $i = 1, ..., r$, then $\theta_e = 0$, hence $\Theta = 0$, i.e., the connection D is flat. A section s of E with $Ds = 0$ is called **parallel** (with respect to D). The reason for this name will become clear in Sect. 2.11 of this chapter. With this terminology in mind, we may say that if E possesses a parallel frame with respect to D in a neighbourhood of every point of M, the connection D is flat. In Sect. 2.11, we shall show that the converse is also true.

Problem 2.7.4 Show that the trivial bundle admits a globally defined frame parallel with respect to the trivial connection.

If X, Y are tangent vectors of M, then

$$\sum_{j=1}^{r}(\theta_{ij}\wedge\theta_{jk})(X,Y)=\sum_{j=1}^{r}[\theta_{ij}(X)\theta_{jk}(Y)-\theta_{ij}(Y)\theta_{jk}(X)],\quad i,k=1,...,r,$$

or in matrix notation

$$(\theta\wedge\theta)(X,Y)=\theta(X)\theta(Y)-\theta(Y)\theta(X),$$

i.e., $(\theta\wedge\theta)(X,Y)=[\theta(X),\theta(Y)]$, the bracket being the commutator of the matrices $\theta(X)$ and $\theta(Y)$. Thus, Cartan's structure equation can be written as

$$\Theta_e=d\theta_e-[\theta_e,\theta_e].$$

Problem 2.7.5 Prove that $d\Theta=[\theta,\Theta]$, where $[\theta,\Theta]=\theta\wedge\Theta-\Theta\wedge\theta$ is the matrix $[\theta,\Theta]_{ik}=\sum_{j=1}^{r}(\theta_{ij}\wedge\Theta_{jk}-\Theta_{ij}\wedge\theta_{jk})$.

This is a version of the so-called second (differential) Bianchi identity.

Remark 2.7.1 In Wells [25], Chap. III, Sect. 1 the matrix of a connection is the transpose of the matrix we are using in these lectures. In this book, if $f=(e_1,...,e_r)$ is a frame of E, the matrix of a connection D is defined as $\theta(D,f)=[\theta_{\rho\sigma}(D,f)]$ where $De_\sigma=\sum_{\rho=1}^{r}\theta_{\rho\sigma}(D,f)\otimes e_\rho$, see Chapter III, identity (1.10). While, in our notations (the same as in Griffiths and Harris [9]), $De_\sigma=\sum_{\rho=1}^{r}\theta_{\sigma\rho}\otimes e_\rho$. Thus, $\theta_{ij}=\theta_{ji}(D,f)$, hence $\theta_{ij}\wedge\theta_{jk}=\theta_{ji}(D,f)\wedge\theta_{kj}(D,f)=-\theta_{kj}(D,f)\wedge\theta_{ji}(D,f)$. Therefore Cartan's structure equation in the notation of the Wells book takes the form $\Theta(D,f)=d\theta(f)+\theta(f)\wedge\theta(f)$ where $\theta(f)=\theta(D,f)$, Chapter III, identity 1.14). Cartan's structure equation is written in the same way in Kobayashi and Nomizu [14], see vol. I, Chap. III, the end of Sect. 1. Note also that in this book, if ω is a connection matrix, $\omega\wedge\omega=\frac{1}{2}[\omega,\omega]$, while the factor $\frac{1}{2}$ is missing in our considerations. This is because the definition of exterior product in Kobayashi and Nomizu [14] differs from the one adopted here (which coincides with the definition in Godbillon [7]). If ω and φ are forms of type p and q, respectively, their exterior product is defined in Kobayashi and Nomizu [14] as $\omega\wedge\varphi=\frac{1}{(p+q)!}$(anti-symmetrization of $\omega\otimes\varphi$), while in these lectures the factor is $\frac{1}{p!q!}$. Because of that, the exterior derivative $d\varphi$ of a differential p-form φ used here is $(p+1)$ times that in Kobayashi and Nomizu

[14], Chap. I, Proposition 3.11: if $X_1, ..., X_{p+1}$ are vector fields

$$(d\varphi)(X_1, ..., X_{p+1}) = \sum_{k=1}^{p+1} (-1)^{k+1} X_k(\varphi(X_1, ..., \widehat{X}_k, ..., X_{p+1})) \\ + \sum_{k<l} (-1)^{k+l} \varphi([X_k, X_l], X_1, ..., \widehat{X}_k, ..., \widehat{X}_l, ..., X_{p+1}).$$

Problem 2.7.6 Let ∇ be a connection on a manifold (i.e., on its tangent bundle) with vanishing torsion: $\nabla_X Y - \nabla_Y X = [X, Y]$ for every vector fields X and Y. Show that

$$(d\varphi)(X_1, ..., X_{p+1}) = \sum_{k=1}^{p+1} (-1)^{k+1} (\nabla_{X_k}\varphi)(X_1, ..., \widehat{X}_k, ..., X_{p+1}),$$

the anti-symmetrization of $\nabla\varphi$.

Proposition 2.7.1 *Let D be a connection on a vector bundle $E \to M$. Denote the induced connection on the bundle $Hom(E, E)$ again by D. Also, let D be the operator on the p-forms on M with values in $Hom(E, E)$ induced by this connection. Then*

$$DR = 0.$$

Proof Let $e_1, ..., e_r$ be a frame of the bundle E. Define sections L_{ij} of the bundle $Hom(E, E)$, $i, j = 1, ..., r$, by $L_{ij}(e_k) = \delta_{ik} e_j$ for $k = 1, ..., r$. These sections constitute a frame of the bundle $Hom(E, E)$. Let $\Theta = [\Theta_{ij}]$ be the curvature matrix of the connection D on this bundle with respect to the frame $\{L_{ij}\}$. Then

$$R = \sum_{i,j=1}^{r} \Theta_{ij} \otimes L_{ij},$$

hence

$$DR = \sum_{i,j=1}^{r} (d\Theta_{ij} \otimes L_{ij} + \Theta_{ij} \wedge DL_{ij}).$$

If $\theta = [\theta_{ij}]$ is the matrix of the connection D on E with respect to the frame $e_1, ..., e_r$,

$$(DL_{ij})(e_k) = DL_{ij}(e_k) - L_{ij}(De_k) = \delta_{ik} De_j - L_{ij}(\sum_{l=1}^{r} \theta_{kl} e_l) \\ = \sum_{l=1}^{r} \delta_{ik}\theta_{jl} e_l - \sum_{l=1}^{r} \theta_{kl}\delta_{il} e_j = \sum_{t=1}^{r} (\delta_{ik}\theta_{jt} - \sum_{l=1}^{r} \delta_{il}\delta_{jt}\theta_{kl}) e_t.$$

Hence

$$DL_{ij} = \sum_{k,t=1}^{r} (\delta_{ik}\theta_{jt} - \sum_{l=1}^{r} \delta_{il}\delta_{jt}\theta_{kl}) \otimes L_{kl}.$$

Then

$$\sum_{i,j=1}^{r} \Theta_{ij} \wedge DL_{ij} = \sum_{i,j=1}^{r} \sum_{k,t=1}^{r} (\delta_{ik}\Theta_{ij} \wedge \theta_{jt} - \sum_{l=1}^{r} \delta_{il}\delta_{jt}\Theta_{ij} \wedge \theta_{kt}) \otimes L_{kt}$$
$$= [\sum_{k,t=1}^{r} \sum_{j=1}^{r} \Theta_{kj} \wedge \theta_{jt} - \sum_{k,t=1}^{r} \sum_{i=1}^{r} \Theta_{it} \wedge \theta_{ki}] \otimes L_{kt}$$
$$= \sum_{k,t=1}^{r} \sum_{j=1}^{r} (\Theta_{kj} \wedge \theta_{jt} - \theta_{kj} \wedge \Theta_{jt}) \otimes L_{kt}.$$

It follows that

$$DR = \sum_{k,t=1}^{r} (d\Theta_{kt} \otimes L_{kt} + \Theta_{kt} \wedge DL_{kt}) = \sum_{k,t=1}^{r} [d\Theta_{kt} - \sum_{j=1}^{r} (\theta_{kj} \wedge \Theta_{jt} - \Theta_{kj} \wedge \theta_{jt}] \otimes L_{kt} = 0$$

by the identity $d\Theta = [\theta, \Theta]$. □

Proposition 2.7.2 *Let D be a connection with vanishing torsion on a manifold M. Consider its curvature tensor R as a section of the bundle $Hom(TM \otimes TM \otimes TM, TM)$, $(u, v, w) \to R(u, v)w$, and denote by D the connection on this bundle induced by the connection on M. Write $R(u, v, w)$ instead of $R(u, v)w$. Then for every $X_1, X_2, X_3, Z \in TM$*

$$(D_{X_1} R)(X_2, X_3, Z) + (D_{X_2} R)(X_3, X_1, Z) + (D_{X_3} R)(X_1, X_2, Z) = 0.$$

Proof As noted in Sect. 2.1, the tangent vectors $X_1, X_2, X_3, Z \in T_pM$ can be extended to vector fields in a neighbourhood of p. Denote these vector fields by the same symbols. Then

$$(D_{X_1} R)(X_2, X_3, Z) =$$
$$D_{X_1} R(X_2, X_3, Z) - R(D_{X_1} X_2, X_3, Z) - R(X_2, D_{X_1} X_3, Z) - R(X_2, X_3, D_{X_1} Z).$$

Consider R as a skew-symmetric 2-form on M with values in the bundle $Hom(TM, TM)$. Let D be the operator of differentiation of such forms determined by the connection on $Hom(TM, TM)$ induced by the connection on TM. By the formula for DR stated in Problem 2.6.2

$$((DR)(X_1, X_2, X_3))(Z)$$
$$= (D_{X_1} R(X_2, X_3))(Z) - (D_{X_2} R(X_1, X_3))(Z) + (D_{X_3} R(X_1, X_2))(Z)$$
$$-(R([X_1, X_2], X_3)(Z) + (R([X_1, X_3], X_2], X_2)(Z) - (R([X_2, X_3], X_1)(Z),$$

where D on the right-hand side means the connection on the bundle $Hom(TM, TM)$ induced by the connection on TM.

We have

$$(D_{X_1} R(X_2, X_3))(Z) = D_{X_1}(R(X_2, X_3))(Z) - R(X_1, X_2)(D_{X_1} Z)$$
$$= D_{X_1} R(X_2, X_3, Z) - R(X_1, X_2, D_{X_1} Z).$$

Also,

$$(R([X_1, X_2], X_3)(Z) = R(D_{X_1}X_2, X_3, Z) - R(D_{X_2}X_1, X_3, Z).$$

Combining the identities above, we see that

$$((DR)(X_1, X_2, X_3))(Z)$$
$$= (D_{X_1}R)(X_2, X_3, Z) + (D_{X_2}R)(X_3, X_1, Z) + (D_{X_3}R)(X_1, X_2, Z).$$

This clearly implies our claim since $DR = 0$ by the preceding proposition. □

Each of the identities stated in Propositions 2.7.1 and 2.7.2 is known as **the second (differential) Bianchi identity**. For a direct proof of Proposition 2.7.2, see, e.g., O'Neill [21], Chap. 3, Proposition 37.

Proposition 2.7.3 *The curvature tensor R of a connection D with vanishing torsion on a manifold satisfies the identity*

$$R(X, Y)Z + R(Y, Z)X + R(Z, X)Y = 0,$$

the first (algebraic) Bianchi identity

Proof If X, Y, Z are vector fields,

$$R(X, Y)Z + R(Y, Z)X + R(Z, X)Y$$
$$= D_X[Y, Z] + D_Z[X, Y] + D_Y[Z, X] - D_{[X,Y]}Z - D_{[Y,Z]}X - D_{[Z,X]}Y$$
$$= [X, [Y, Z]] + [Y, [Z, X]] + [Z, [X, Y]] = 0$$

by the Jacobi identity. □

Now, let $\varphi \in A^p(E)$ and set $\varphi = \sum_{i=1}^r \varphi_i \otimes e_i$. Then, by (2.9),

$$\begin{aligned} D^2\varphi &= D(\sum_{j=1}^r [d\varphi_j + (-1)^p \sum_{i=1}^r \varphi_i \wedge \theta_{ij}] \otimes e_j) \\ &= \sum_{k=1}^r [d(d\varphi_k + (-1)^p \sum_{i=1}^r \varphi_i \wedge \theta_{ik}) \\ &\qquad + (-1)^{p+1} \sum_{j=1}^r (d\varphi_j + (-1)^p \sum_{i=1}^r \varphi_i \wedge \theta_{ij}) \wedge \theta_{jk}] \otimes e_k \\ &= \sum_{i,k=1}^r [\varphi_i \wedge (d\theta_{ik} - \sum_{j=1}^r \theta_{ij} \wedge \theta_{jk})] \otimes e_k \\ &= \sum_{i,k=1}^r [\varphi_i \wedge \Theta_{ik}] \otimes e_k = \sum_{i=1}^r \varphi_i \wedge (\sum_{k=1}^r \Theta_{ik} \otimes e_k) = \sum_{i=1}^r \varphi_i \wedge D^2 e_i. \end{aligned}$$

The sum $\sum_{i=1}^{r} \varphi_i \wedge D^2 e_i = \sum_{i,k=1}^{r} [\varphi_i \wedge \Theta_{ik}] \otimes e_k$ does not depend on the choice of the frame $e_1, ..., e_r$ and is denoted by $\varphi \wedge R$. Under this notation,

$$D^2\varphi = \varphi \wedge R.$$

Let σ be a section of E over an open set U, and let X, Y be vector fields on U. Also, let $e_1, ..., e_r$ be a frame of E over U. If $\sigma = \sum_{i=1}^{r} \sigma_i e_i$, then

$$\begin{aligned}(D^2\sigma)(X, Y) &= \sum_{i=1}^{r} \sigma_i (D^2 e_i)(X, Y) = \sum_{i,k=1}^{r} \sigma_i (d\theta_{ik} - \sum_{j=1}^{r} \theta_{ij} \wedge \theta_{jk})(X, Y) e_k \\ &= \sum_{i,k=1}^{r} \sigma_i (X(\theta_{ik}(Y)) - Y(\theta_{ik}(X)) - \theta_{ik}([X, Y])) e_k \\ &\quad - \sum_{i,j,k=1}^{r} \sigma_i (\theta_{ij}(X)\theta_{jk}(Y) - \theta_{ij}(Y)\theta_{jk}(X)) e_k.\end{aligned}$$

Moreover,

$$D_Y\sigma = \sum_{i=1}^{r} Y(\sigma_i) e_i + \sum_{i,j=1}^{r} \sigma_i \theta_{ij}(Y) e_j.$$

Hence

$$\begin{aligned}D_X D_Y \sigma = &\sum_{k=1}^{r} XY(\sigma_k) e_k + \sum_{i,k=1}^{r} Y(\sigma_i)\theta_{ik}(X) e_k \\ &+ \sum_{i,k=1}^{r} X(\sigma_i)\theta_{ik}(Y) e_k + \sum_{i,k=1}^{r} \sigma_i X(\theta_{ik}(Y)) e_k \\ &+ \sum_{i,j,k=1}^{r} \sigma_i \theta_{ij}(Y)\theta_{jk}(X) e_k.\end{aligned}$$

It follows that

$$(D^2\sigma)(X, Y) = D_X D_Y \sigma - D_Y D_X \sigma - D_{[X,Y]}\sigma.$$

Therefore

$$R(X, Y)\sigma = D_X D_Y \sigma - D_Y D_X \sigma - D_{[X,Y]}\sigma.$$

Remark 2.7.2 Sometimes, it is more convenient to define the curvature tensor as $-D_X D_Y \sigma + D_Y D_X \sigma + D_{[X,Y]}\sigma$.

By the formula above

$$D_X D_Y e_i - D_Y D_X e_i - D_{[X,Y]} e_i = R(X, Y) e_i = \sum_{j=1}^{r} \Theta_{ij}(X, Y) e_j, \quad i = 1, ..., r. \tag{2.10}$$

Problem 2.7.7 Let $E \to M$ be a vector bundle endowed with a metric and a metric connection. Denote the curvature tensor of this connection by R. Prove that:

(i) In the case when E is a real bundle with an Euclidean metric g,

$$g(R(X,Y)s,\sigma) = -g(R(X,Y)\sigma,s), \quad X,Y \in T_xM,\ s,\sigma \in E_x;$$

(ii) In the case when E is a complex bundle with a Hermitian metric h,

$$h(R(Z,W)s,\sigma) = -h(s,R(\overline{Z},\overline{W})\sigma), \quad Z,W \in T_x^{\mathbb{C}}M,\ s,\sigma \in E_x.$$

It follows from (i) that the curvature tensor R of a metric connection on E can be considered as a 2-form on M with values in the bundle $\mathcal{A}(E,E)$ of skew-symmetric endomorphisms of E.

Note also that identity (i) and the algebraic Bianchi identity imply that if R is the curvature tensor of the Levi-Civita connection of a Riemannian manifold (M,g),

$$g(R(X,Y)Z,U) = g(R(Z,U)X,Y).$$

Problem 2.7.8 Prove the latter identity.

Let D be a metric connection on a Riemannian manifold (M,g). In accordance with Sect. 2.4, denote the metric on the bundle Λ^2TM induced by the metric g on TM by g^{Λ}. The identities $g(R(X,Y)Z,U) = -g(R(Y,X)Z,U)$ and $g(R(X,Y)Z,U) = -g(R(X,Y)U,Z)$ allow one to define a linear operator $\mathcal{R} : \Lambda^2TM \to \Lambda^2TM$ setting

$$g^{\Lambda}(\mathcal{R}(X\wedge Y), Z\wedge U) = g(R(X,Y)Z,U).$$

If D is the Levi-Civita connection, this operator is symmetric: $g^{\Lambda}(\mathcal{R}(X\wedge Y), Z\wedge U) = g^{\Lambda}(X\wedge Y, \mathcal{R}(Z\wedge U))$. The operator $\mathcal{R}$ plays an important role in studying the curvature.

Problem 2.7.9 Let $E \to M$ and $F \to M$ be vector bundles endowed with connections D^E, D^F and corresponding curvature tensors R^E, R^F. Consider the vector bundles $E\oplus F$, $E\otimes F$, E^*, $Hom(E,F)$, Λ^pE, $\odot^pE$ with the connections induced by D^E and D^F. Show that the corresponding curvature tensors satisfy the following identities.

(1) If $s \in E_x$, $\sigma \in F_x$ and $X,Y \in T_xM$, then

$$R^{E\oplus F}(X,Y)(s+\sigma) = R^E(X,Y)s + R^F(X,Y)\sigma.$$

(2) $R^{E\otimes F}(X,Y)(s\otimes\sigma) = R^E(X,Y)s\otimes\sigma + s\otimes R^F(X,Y)\sigma$.

(3) If $\alpha \in E_x^*$,

$$(R^{E^*}(X,Y)\alpha)(s) = -\alpha(R^E(X,Y)s).$$

(4) If $\varphi \in Hom(E,F)_x$,

$$(R(X,Y)\varphi)(s) = R^F(X,Y)\varphi(s) - \varphi(R^E(X,Y)s).$$

(5) If $s_1, ..., s_p \in E_x$, then

$$R(X, Y)(s_1 \wedge ... \wedge s_p) = R^E(X, Y)s_1 \wedge ... \wedge s_p + \cdots + s_1 \wedge ... \wedge R^E(X, Y)s_p.$$

(6) $R(X, Y)(s_1 \odot ... \odot s_p) = R^E(X, Y)s_1 \odot ... \odot s_p + \cdots + s_1 \odot ... \odot R^E(X, Y)s_p.$
(7) If $f : \widetilde{M} \to M$ is a smooth map, let $\widetilde{D}$ be the connection on the pull-back bundle $\widetilde{E} = f^*E$ induced by D^E. Denote by $\widetilde{f} : \widetilde{E} \to E$ the standard map $\widetilde{E}_x \ni (x, v) \to v \in E_{f(x)}$. Then, if $\widetilde{s} \in \widetilde{E}_x$ and $X, Y \in T_x\widetilde{M}$, the curvature $\widetilde{R}$ of $\widetilde{D}$ satisfies the identity

$$\widetilde{R}(X, Y)\widetilde{s} = R^E(f_*X, f_*Y)\widetilde{f}(\widetilde{s}).$$

Now, let $\pi : E \to M$ be a holomorphic vector bundle over a complex manifold M. Let D be a connection on E consistent with the complex structure of E. Then, as we have mentioned, the matrix $[\theta_{ij}]$ of the connection D with respect to any holomorphic frame is of type $(1, 0)$. It follows from Cartan's structure equation $\Theta_{ik} = d\theta_{ik} - \sum_{j=1}^r \theta_{ij} \wedge \theta_{jk}$ that Θ_{ik} is the sum of forms of type $(2, 0)$ and $(1, 1)$, i.e., the $(0, 2)$-part of Θ_{ik} vanishes , $\Theta_{ik}^{0,2} = 0$. It follows from the formula for changing the curvature matrix when changing the frame that the type of the matrix $\Theta = [\Theta_{ik}]$ does not change when changing the frame. Thus, $\Theta^{0,2} = 0$ with respect to every frame of E.

Suppose further that E possesses a Hermitian metric h and D is consistent with the complex structure and the metric on E. Let $e = (e_1, ..., e_r)$ be a holomorphic frame of E, and set $h = [h(e_i, e_j)]$. Then, as we know from the proof of Proposition 2.5.1, the matrix of the connection D with respect to the frame e is $\theta = \partial h \cdot h^{-1}$. It follows from the identity $hh^{-1} = I_r$ (the identity $r \times r$-matrix) that $\partial h \cdot h^{-1} + h \cdot \partial h^{-1} = 0$, hence $\partial h^{-1} = -h^{-1} \cdot \partial h \cdot h^{-1}$. Then, taking into account the identity $\partial^2 = 0$, we get

$$\partial\theta = \partial^2 h \cdot h^{-1} - \partial h \wedge (-h^{-1} \cdot \partial h \cdot h^{-1}) = (\partial h \cdot h^{-1}) \wedge (\partial h \cdot h^{-1}) = \theta \wedge \theta.$$

Therefore

$$\Theta = \partial\theta + \overline{\partial}\theta - \theta \wedge \theta = \overline{\partial}\theta.$$

In particular, $\overline{\partial}\Theta = 0$. Furthermore, we see that Θ is of type $(1, 1)$ since θ is of type $(1, 0)$. It follows from the formula for changing the curvature matrix when changing the frame that Θ is of type $(1, 1)$ with respect to any frame of E. The matrix of the connection D with respect to any orthonormal frame is skew-Hermitian, $\theta_{ij} + \overline{\theta}_{ji} = 0$, since D is compatible with the metric h. Cartan's structure equation implies that $\Theta_{ij} + \overline{\Theta}_{ji} = 0$. Therefore the curvature matrix of the canonical connection on a Hermitian holomorphic vector bundle is skew-Hermitian and of type $(1, 1)$.

Let $e_1, ..., e_r$ be an orthonormal (unitary) frame, and let $\sigma = \sum_{i=1}^r \sigma_i e_i$ be a section of E. The identity $R(v, w)\sigma = \sum_{i,j=1}^r \sigma_i \Theta_{ij}(v, w)e_j$, $v, w \in T^{\mathbb{C}}M$, shows that the form $R \in A^2(Hom(E, E))$ is of type $(1, 1)$.

Example 2.7.1 Let $L \to M$ be a Hermitian holomorphic line bundle with metric h. Let $\{U_\alpha, \varphi_\alpha\}$ be an atlas of (holomorphic) trivializations of L. Denote the transition functions of this atlas by $g_{\alpha\beta} : U_\alpha \cap U_\beta \to \mathbb{C}\setminus\{0\}$. Every section $e_\alpha(x) = \varphi_\alpha^{-1}(x, 1)$ is a holomorphic frame of L on U_α. Set $h_\alpha = h(e_\alpha, e_\alpha)$. The functions h_α are smooth and positive on U_α. We have $h_\alpha = |g_{\alpha\beta}|^2 h_\beta$ since $e_\alpha = g_{\alpha\beta} e_\beta$. The matrix of the Chern connection on the bundle L consists of one element, the 1-form $\theta_\alpha = h_\alpha^{-1}\partial h_\alpha = \partial \ln h_\alpha$. Then the curvature matrix Θ_α of the Chern connection is the $(1, 1)$-form

$$\Theta_\alpha = \overline{\partial}\theta_\alpha = \overline{\partial}\partial \ln h_\alpha = -\partial\overline{\partial} \ln h_\alpha = -h_\alpha^{-2}\partial h_\alpha \wedge \overline{\partial} h_\alpha.$$

The connection forms θ_α and θ_β are related by $\theta_\alpha = |g_{\alpha\beta}|^{-2} d\, |g_{\alpha\beta}|^2 + \theta_\beta$. Since the functions $g_{\alpha\beta}$ are holomorphic on $U_\alpha \cap U_\beta$,

$$\theta_\alpha = (g_{\alpha\beta}\overline{g_{\alpha\beta}})^{-1}\partial g_{\alpha\beta}\overline{g_{\alpha\beta}} + \theta_\beta = g_{\alpha\beta}^{-1}\partial g_{\alpha\beta} + \theta_\beta.$$

Therefore on $U_\alpha \cap U_\beta$

$$\Theta_\alpha = \overline{\partial}\theta_\alpha = \Theta_\alpha = \overline{\partial}\theta_\beta = \Theta_\beta.$$

This identity shows that the locally defined forms Θ_α determine a globally defined $(1, 1)$-form Θ on M, the curvature form of the Chern connection.

Now, let M be a Hermitian complex manifold and D the canonical connection on the holomorphic tangent bundle $T^h M$. Recal that as a smooth complex vector bundle $T^h M$ is isomorphic to $T^{1,0}M$, hence D can be considered as a connection on the bundle $T^{1,0}M$. Let $Z_1, \ldots, Z_n$ be an orthonormal (unitary) frame of $(1, 0)$ vector fields and $\varphi_1, \ldots, \varphi_n$ its dual frame of $(1, 0)$-forms. Let D^* be the connection on $(T^{1,0}M)^*$ yielded by D. Denote by θ and θ^* the matrices of the connections D and D^* with respect to the frames $Z_1, \ldots, Z_n$ and $\varphi_1, .., \varphi_n$. Then $\theta = -{}^t\theta^*$. Since D^* is compatible with the metric of $(T^{1,0}M)^*$ and the frame $\varphi_1, \ldots, \varphi_n$ is orthonormal, we have

$(a) \quad \theta^* + {}^t\overline{\theta^*} = 0.$

Moreover, the connection D^* is consistent with the complex structure, hence $D^{*\prime\prime} = \overline{\partial}$. For $D^{*\prime\prime}$, we have $D^{*\prime\prime}\varphi_i = \sum_{j=1}^r (\theta_{ij}^*)^{0,1} \otimes \varphi_j$. Therefore

$(b) \quad \overline{\partial}\varphi_i = \sum_{j=1}^r (\theta_{ij}^*)^{0,1} \wedge \varphi_j.$

The next lemma shows that the identities (a) and (b) satisfied by θ^* uniquely determine this matrix.

Lemma 2.7.1 *Let M be a complex manifold and $\varphi_1, \ldots, \varphi_n$ a frame of $(1, 0)$-forms on M. Then there exists a unique matrix $\psi = [\psi_{ij}]$ of 1-forms such that*

(i) $\psi + {}^t\overline{\psi} = 0$;
(ii) $\overline{\partial}\varphi_i = \sum_{j=1}^n \psi_{ij}^{0,1} \wedge \varphi_j$.

Proof Suppose there exists a matrix ψ with the properties (i) and (ii). Since $\overline{\varphi}_1, ..., \overline{\varphi}_n$ is a frame of $(0, 1)$-forms, $\psi_{ij}^{0,1} = \sum_{k=1}^n f_{ij}^k \overline{\varphi}_k$ where f_{ij}^k are smooth functions. Then, by (ii), $\overline{\partial}\varphi_i = \sum_{j,k=1}^n f_{ij}^k \overline{\varphi}_k \wedge \varphi_j$. Since the forms $\overline{\varphi}_k \wedge \varphi_j$ constitute a frame, the latter identity uniquely determines the functions f_{ij}^k, hence $\psi_{ij}^{0,1}$. It follows from (i) that $\psi_{ij}^{1,0} = -\overline{\psi_{ji}^{0,1}}$. This shows that the forms ψ_{ij} are uniquely determined. The reasoning above shows also the existence of a matrix ψ with the desired properties. □

According to Lemma 2.7.1, in order to compute the matrix θ^* of the connection D^*, it suffices to represent $\overline{\partial}\varphi_i$ in terms of the frame $\overline{\varphi}_k \wedge \varphi_j$ of $(1, 1)$-forms. If $\overline{\partial}\varphi_i = \sum_{k=1}^n f_{ij}^k \overline{\varphi}_k \wedge \varphi_j$, then $(\theta_{ij}^*)^{0,1} = \sum_{k=1}^n f_{ij}^k \overline{\varphi}_k$ and $(\theta_{ij}^*)^{1,0} = -\overline{(\theta_{ji}^*)^{0,1}} = -\sum_{k=1}^n \overline{f_{ji}^k} \varphi_k$. In this way, we find $\theta_{ij}^* = (\theta_{ij}^*)^{1,0} + (\theta_{ij}^*)^{0,1}$ and $\theta_{ij} = -\theta_{ji}^*$.

The point of these considerations is that the computation of the curvature matrix in the case of a Hermitian manifold can be reduced to exterior differentiation.

Example 2.7.2 We apply the method above to compute the matrix of the canonical connection of a Riemann surface M endowed with a Hermitian metric h.

Let z be a local coordinate of M. Then $h = F dz \otimes d\overline{z}$ where $F = h(\frac{\partial}{\partial z}, \frac{\partial}{\partial \overline{z}}) > 0$ is a smooth function. Set $f = \sqrt{F}$ and write h in the form $h = f^2 dz \otimes d\overline{z}$. The 1-form $\varphi = f dz$ is an orthonormal frame for the $(1, 0)$-forms. We have

$$\overline{\partial}\varphi = \overline{\partial} f \wedge dz = \frac{\overline{\partial} f}{f} \wedge \varphi = \overline{\partial}\, ln\, f \wedge \varphi.$$

Therefore $(\theta^*)^{0,1} = \overline{\partial}\, ln\, f$, hence $(\theta^*)^{1,0} = -\overline{(\theta^*)^{0,1}} = -\partial\, ln\, f$. Thus, $\theta^* = (\overline{\partial} - \partial) ln\, f$, hence

$$\theta = -\theta^* = (\partial - \overline{\partial}) ln\, f = \frac{\partial\, ln\, f}{\partial z} dz - \frac{\partial\, ln\, f}{\partial \overline{z}} d\overline{z}.$$

Since $\theta \wedge \theta = 0$ (θ is a 1-form), by the Cartan structure equation

$$\Theta = d\theta = \frac{\partial^2 ln\, f}{\partial \overline{z} \partial z} d\overline{z} \wedge dz - \frac{\partial^2 ln\, f}{\partial z \partial \overline{z}} dz \wedge d\overline{z} = -2 \frac{\partial^2 ln\, f}{\partial z \partial \overline{z}} dz \wedge d\overline{z}.$$

Let $z = x + iy$ and $\Delta = \frac{\partial}{\partial x^2} + \frac{\partial}{\partial y^2}$. Then $\frac{\partial}{\partial z \partial \overline{z}} = \frac{1}{4}\Delta$ and

$$\Theta = -\frac{1}{2}(\Delta\, ln\, f) dz \wedge d\overline{z}.$$

The Gauss curvature of the metric $g = Re\, h = f^2(dx^2 + dy^2)$ is

$$K = -\frac{1}{2}\frac{\Delta \ln f^2}{f^2},$$

so

$$\Theta = \frac{f^2}{2} K dz \wedge d\overline{z}.$$

If $v \in T^{1,0}M$,

$$\Theta(v, \overline{v}) = \frac{1}{2} K |v|_h^2 \quad (|v|_h \text{ is the norm of } v \text{ with respect to the metric } h).$$

Hence $\Theta(v, \overline{v}) \gtreqless 0$ for $v \neq 0$ if and only if $K \gtreqless 0$.

Recall that by Lemma 2.7.1, for every frame $\varphi_1, ..., \varphi_n$ of $(1, 0)$-forms on a complex manifold M, there exist uniquely determined 1-forms ψ_{ij} such that (i) $\psi_{ij} + \overline{\psi}_{ji} = 0$; (ii) $\overline{\partial}\varphi_i = \sum_{j=1}^n \psi_{ij}^{0,1} \wedge \varphi_j$. If identity (ii) holds, then

$$(ii') \quad d\varphi_i = \sum_{j=1}^n \psi_{ij} \wedge \varphi_j + \tau_i,$$

where $\tau_i = -\sum_{j=1}^n \psi_{ij}^{1,0} \wedge \varphi_j + \partial\varphi_i$ are $(2, 0)$-forms. Conversely, if the identity (ii') holds for some $(2, 0)$-forms τ_i, it is clear that identity (ii) holds. Hence the forms ψ_{ij} and τ_i are uniquely determined by identities (i) and (ii), as well as identities (i) and (ii').

The set of forms $\tau_1, ..., \tau_n$ is sometimes referred to as the torsion of the complex structure with respect to the frame $\varphi_1, ..., \varphi_n$. The reason for this is the following.

Let M be a Hermitian manifold and, as above, $Z_1, ..., Z_n$ an orthonormal (unitary) frame of $(1, 0)$ vector fields. Let $\varphi_1, ..., \varphi_n$ be the dual frame of $(1, 0)$-forms. Denote the canonical connection on $T^h M \cong T^{1,0}M$ by D. Then the induced connection D^* is the canonical connection on $(T^{1,0}M)^*$. Denote by θ and θ^* the connection matrices of D and D^* with respect to the frames $Z_1, ..., Z_n$ and $\varphi_1, ..., \varphi_n$. Taking into account the identities (a) and (b) above and the fact that $\theta^* = -{}^t\theta$, we see that the forms τ_i determined via the frame $\varphi_1, ..., \varphi_n$ satisfy the identity $\tau_i = \sum_{j=1}^n \theta_{ji}^{1,0} \wedge \varphi_j + \partial\varphi_i$. Then if V and W are $(1, 0)$-vector fields on M,

$$\tau_i(V, W) = \sum_{j=1}^n [\theta_{ji}(V)\varphi_j(W) - \theta_{ji}(W)\varphi_j(V)] + V(\varphi_i(W)) - W(\varphi_i(V)) - \varphi_i([V, W]).$$

Moreover $W = \sum_{k=1}^n \varphi_k(W) Z_k$, hence

$$D_V W = \sum_{k=1}^n V(\varphi_k(W)) Z_k + \sum_{k,l=1}^n \varphi_k(W)\theta_{kl}(V) Z_l.$$

We get from the latter identity that

$$\varphi_i(D_V W) = V(\varphi_i(W)) + \sum_{k=1}^{n} \varphi_k(W)\theta_{ki}(V).$$

It follows that

$$\tau_i(V, W) = \varphi_i(D_V W - D_W V - [V, W]) = \varphi_i(T(V, W)),$$

where T is the torsion of the canonical connection of M. This implies $T(V, W) = \sum_{i=1}^{n} \tau_i(V, W)Z_i$ since, as we have seen in Sect. 2.5, $T(V, W)$ is of type $(1, 0)$. In Sect. 2.5, we also noted that the tensor T is uniquely determined by its values $T(V, W)$. The latter identity therefore justifies the name "torsion forms" for the forms τ_i. This identity implies that $\tau_i(V, W) = 0$ for every $i = 1, ..., n$ if and only if the torsion of the canonical connection vanishes. As we have observed in Sect. 2.5, the torsion of the canonical connection of a Hermitian manifold vanishes if and only if the manifold is Kähler.

Problem 2.7.10 Let M be a Hermitian complex manifold such that around each of its points there exists an orthonormal frame of $(1, 0)$-forms for which the torsion of the complex structure vanishes, $\tau_i = 0$, $i = 1, ..., n$. Show by a direct computation that the fundamental 2-form of M is closed.

2.8 The Second Fundamental Form of a Hermitian Holomorphic Subbundle and its Curvature

Let $\pi : E \to M$ be a Hermitian holomorphic vector bundle with a metric h. Let F be a holomorphic subbundle of E. Then the quotient bundle E/F is a holomorphic vector bundle and the natural projection $p : E \to E/F$ is a holomorphic map, a bundle morphism. The orthogonal complement of F,

$$F^{\perp} = \{e \in E : h(e, f) = 0 \text{ for every } f \in E_{\pi(e)}\}$$

is a smooth subbundle of E, but it may not be a holomorphic subbundle (this is the case when F is the tautological bundle, see Chap. 3). To show that $F^{\perp}$ is a smooth subbundle of E, it suffices to take a frame $s_1, ..., s_r$ of E such that $s_1, ..., s_m$ is a frame of F. Applying the Gram-Schmidt orthogonalization process, we obtain a frame $\widetilde{s}_1, ..., \widetilde{s}_r$ of E such that the values of $\widetilde{s}_{m+1}, ..., \widetilde{s}_r$ lie in $F^{\perp}$ (and $\widetilde{s}_1, ..., \widetilde{s}_m$ is a frame of F). Then, by Chap. 1, Proposition 1.7.3, $F^{\perp}$ possesses a unique structure of a smooth submanifold of E such that $\widetilde{s}_{m+1}, ..., \widetilde{s}_r$ is a frame of this subbundle.

Let $p_{F^{\perp}} : E \to F^{\perp}$ be the orthogonal projection. As smooth bundles, $F^{\perp}$ and E/F are isomorphic via the map $p|F^{\perp} : F^{\perp} \to E/F$; the inverse map sends $p(e)$, $e \in E$, to $p_{F^{\perp}}(e)$. Transfer to $F^{\perp}$ the structure of the complex manifold E/F by means of the diffeomorphism $p|F^{\perp}$. In this way, we obtain a holomorphic bundle whose underlying smooth bundle is $F^{\perp}$. Denote this holomorphic bundle by Q (it is isomorphic to E/F). Consider the bundle Q with the Hermitian metric $h|F^{\perp}$. In order to compute the canonical connection of the Hermitian holomorphic vector bundle Q, we find the action of the $\overline{\partial}$-operator on the smooth sections of Q. Denote this operator by $\overline{\partial}_Q$, and let $\overline{\partial}_E$ be the $\overline{\partial}$-operator of E. Let σ be a smooth section of Q, i.e., of $F^{\perp}$. Since F is a holomorphic subbundle of E, there exists a holomorphic frame $e_1, ..., e_r$ of E such that $e_1, ..., e_m$ is a holomorphic frame of F. Then $p(e_{m+1}), ..., p(e_r)$ is a holomorphic frame of E/F, hence $p_{F^{\perp}}(e_{m+1}), ..., p_{F^{\perp}}(e_r)$ is a holomorphic frame of Q. We can consider every smooth section σ of Q as a smooth section of E since Q is a smooth subbundle of E. Thus, we can represent σ in the form $\sigma = \sum_{i=1}^{r} \sigma_i e_i$ where σ_i are smooth functions. Then $\sigma = p_{F^{\perp}}(\sigma) = \sum_{l=m+1}^{r} \sigma_l p_{F^{\perp}}(e_l)$ as a section of Q. Therefore, if $w \in T^{0,1}M$,

$$(\overline{\partial}_Q\sigma)(w) = \sum_{l=m+1}^{r} w(\sigma_l)p_{F^{\perp}}(e_l) = p_{F^{\perp}}(\sum_{l=1}^{r} w(\sigma_l)e_l) = p_{F^{\perp}}((\overline{\partial}_E\sigma)(w)).$$

Thus, $\overline{\partial}_Q = p_{F^{\perp}} \circ \overline{\partial}_E$. It follows that if D^E is the canonical connection of E, the connection $p_{F^{\perp}} \circ D^E|A^0(Q)$ is consistent with the complex structure of Q. It is easy to check that this connection is compatible with the metric $h|F^{\perp}$ on Q. Therefore the canonical connection of Q is $D^Q = p_{F^{\perp}} \circ D^E|A^0(Q)$.

Recall that the canonical connection of F is $D^F = p_F \circ D^E|A^0(F)$ where $p_F : E \to F$ is the orthogonal projection. If $F^{\perp}$ is a holomorphic subbundle of E, then $E \cong F \oplus F^{\perp}$ and $D^E = D^F \oplus D^{F^{\perp}}$, hence $D^E|A^0(F) - D^F = 0$. Therefore we can consider the difference $D^E|A^0(F) - D^F$ as a measure of the extent to which $F^{\perp}$ is not a holomorphic subbundle of E.

Definition 2.8.1 The operator

$$A = D^E|A^0(F) - D^F = p_{F^{\perp}} \circ D^E|A^0(F) : A^0(F) \to A^1(Q)$$

is called the second fundamental form of the holomorphic subbundle F (the first fundamental form of F is its metric $h|F$).

Let $\sigma \in A^0(F)$ and let f be a smooth (local) function on M. Then

$$A(f\sigma) = df \otimes \sigma + fD^E\sigma - df \otimes \sigma - fD^F\sigma = fA(\sigma).$$

Hence A determines a smooth section of the bundle $(T^*M)^{\mathbb{C}} \otimes Hom(F, Q)$. Its value for $\xi \in F_x$, $x \in M$, and $w \in T_x^{\mathbb{C}}M$ is defined by taking a section σ of F with $\sigma(x) = \xi$ and setting $A_w(\xi) = (A_w\sigma)(x)$. Let $w \in T^{0,1}M$. If $e_1, ..., e_m$ is a holomorphic frame of F in a neighbourhood of the point x and $\sigma = \sum_{k=1}^{m} \sigma_k e_k$, then

$$A_w(\sigma) = \sum_{k=1}^{m} \sigma_k A_w(e_k) = \sum_{k=1}^{m} \sigma_k(D_w^E e_k - D_w^F e_k) = 0.$$

Therefore A is of type $(1, 0)$, i.e., $A \in A^{1,0}(Hom(F, Q))$

Denote the curvature tensor of the connection D^E by R^E. Let $e_1, ..., e_r$ be an orthonormal frame of E. For every two vectors $v, w \in T_x^{\mathbb{C}} M$ and every two sections $\sigma = \sum_{i=1}^r \sigma_i e_i$ and $s = \sum_{i=1}^r s_i e_i$ of E in a neighbourhood of x,

$$h(R^E(v, w)\sigma, s) = \sum_{i,j=1}^{r} \Theta_{ij}(v, w)\sigma_i \overline{s}_j,$$

where $[\Theta_{ij}]$ is the curvature matrix of D^E with respect to the frame $e_1, ..., e_r$. This matrix is of type $(1, 1)$ and is skew-Hermitian. Note that we have for $v, w \in T_x^{1,0} M$

$$\Theta_{ij}(v, \overline{w}) = -\overline{\Theta}_{ji}(v, \overline{w}) = -\overline{\Theta_{ji}(\overline{v}, w)} = \overline{\Theta_{ji}(w, \overline{v})}.$$

In particular, $\Theta_{ij}(v, \overline{v}) = \overline{\Theta_{ji}(v, \overline{v})}$. This shows that the form $h(R^E(v, \overline{v})\sigma, \sigma) = \sum_{i,j=1}^{r} \Theta_{ij}(v, \overline{v})\sigma_i \overline{\sigma}_j$ is Hermitian.

Definition 2.8.2 The real number $K^E(v; \sigma) = h(R^E(v, \overline{v})\sigma, \sigma)$, $v \in T_x^{1,0} M$, $\sigma \in E_x$, is called the curvature of E determined by $\{v, \sigma\}$.

If M is a complex manifold and $E = T^h M$, the curvature of E will be denoted by K^M. As a smooth complex bundle, $E = T^h M$ is isomorphic to $T^{1,0} M$, and the connection D^E can be considered as a connection on the bundle $T^{1,0} M$. Let $g = Re\, h$. The Hermitian form h can be viewed as a $\mathbb{C}$-bilinear form on $T^{1,0} M \times T^{0,1} M$ via the standard isomorphisms $E = T^h M \cong T^{1,0} M$ and $\overline{E} \cong T^{0,1} M$. Extend the Euclidean metric g to $T^{\mathbb{C}} M$ by complex bilinearity. Then, as we have noted in Sect. 2.4, $h = 2g|(T^{1,0} M \times T^{0,1} M)$. Hence, for $v = X - iJX$, $w = Y - iJY \in T_x^{1,0} M$, we have $K(v; w) = 2g(R(v, \overline{v})w, \overline{w})$. Making use of Chap. 2, Problem 2.7.7, identity (i) for the curvature tensor, we easily obtain $\frac{1}{8} K(v; w) = g(R^E(JX, X)Y, JY)$. This number is called the **holomorphic bisectional curvature** determined by the pair $\{X, Y\}$. It was coined by Goldberg and Kobayashi in 1967 for Kähler manifolds (recall that in this case, the canonical connection of h coincides with the Levi-Civita connection of g). If $X = Y$, the holomorphic bisectional curvature is called the **holomorphic sectional curvature**.

For $v \in T_x^{\mathbb{C}} M$, denote by $(A_v)^* : Q_x \to F_x$ the adjoint operator of the operator A_v with respect to the metric h, so $(A_v)^*$ is determined by the identity $h((A_v)^*(\zeta), \xi) = h(\zeta, A_v(\xi))$, $\zeta \in Q_x$, $\xi \in F_x$. Since $(A_v)^*$ is anti-linear in v, the map $A^* : T_x^{\mathbb{C}} M \ni v \to (A_{\overline{v}})^* \in Hom(Q_x, F_x)$ is $\mathbb{C}$-linear and defines a 1-form with values in the bundle $Hom(Q, F)$. In fact, $A^* \in A^{0,1}(Hom(Q, F))$ since $A_{\overline{v}} = 0$ for every $v \in T^{1,0} M$. Under this notation, we have the following.

Proposition 2.8.1 *Let E be a Hermitian holomorphic vector bundle with a metric h and F a holomorphic subbundle. Denote by R^E, R^F, and R^Q the curvature tensors of the canonical connections D^E, D^F, and D^Q of the Hermitian holomorphic bundles*

E, F, and $Q \cong E/F$ ($Q = F^{\perp}$ as smooth bundles). Let A be the second fundamental form of F. Then for $v, w \in T_x^{1,0}M$, $\sigma, s \in F_x$ and $\varphi, \psi \in Q_x$:

(1) $h(R^E(v,\overline{w})\sigma, s) = h(R^F(v,\overline{w})\sigma, s) + h(A_v(\sigma), A_w(s))$

(2) $h(R^E(v,\overline{w})\varphi, \psi) = h(R^Q(v,\overline{w})\varphi, \psi) - h((A_w)^*(\varphi), (A_v)^*(\psi))$.

Proof Let $e = (e_1, ..., e_r)$ be an orthonormal frame of E such that $e' = (e_1, ..., e_m)$ is a frame of F and $e'' = (e_{m+1}, ..., e_r)$ is a frame of $F^{\perp}$. Let $\theta^E, \theta^F, \theta^Q$ be the matrices of the corresponding canonical connections with respect to the frames e, e', e''. Set

$$Ae_\alpha = \sum_{\varepsilon=m+1}^{r} a_{\alpha\varepsilon} \otimes e_\varepsilon, \quad \alpha = 1, ..., m.$$

Here $a_{\alpha\varepsilon}$ are 1-forms of type $(1,0)$ since, as we have noted, A is of type $(1,0)$. Denote the matrix $[a_{\alpha\varepsilon}]$ again by A. For every $\alpha = 1, ..., m$, we have $Ae_\alpha = p_{F^{\perp}}(D^E e_\alpha) = \sum_{\varepsilon=m+1}^{r} \theta^E_{\alpha\varepsilon} e_\varepsilon$. Since $\theta^E + {}^t\overline{\theta^E} = 0$ with respect to an orthonormal frame, the block-matrix form of θ^E is

$$\theta^E = \begin{bmatrix} \theta^F & A \\ -{}^t\overline{A} & \theta^Q \end{bmatrix} \tag{2.11}$$

(where the matrix ${}^t\overline{A}$ represents the adjoint operator A^* with respect to the frames e'' of Q and e' of F).

By Cartan's structure equation, for $\alpha, \beta = 1, ..., m$,

$$\Theta^E_{\alpha\beta} = d\theta^E_{\alpha\beta} - \sum_{l=1}^{r} \theta^E_{\alpha l} \wedge \theta^E_{l\beta} = d\theta^E_{\alpha\beta} - \sum_{k=1}^{m} \theta^E_{\alpha k} \wedge \theta^E_{k\beta} + \sum_{\varepsilon=m+1}^{r} \theta^E_{\alpha\varepsilon} \wedge \overline{\theta}^E_{\beta\varepsilon}$$
$$= \Theta^F_{\alpha\beta} + \sum_{\varepsilon=m+1}^{r} a_{\alpha\varepsilon} \wedge \overline{a}_{\beta\varepsilon}.$$

Let $\sigma = \sum_{k=1}^{m} \sigma_k e_k$ and $s = \sum_{k=1}^{m} s_k e_k$. Then

$$h(R^E(v,\overline{w})\sigma, s) = \sum_{\alpha,\beta=1}^{m} \Theta^E_{\alpha\beta}(v,\overline{w})\sigma_\alpha \overline{s}_\beta$$
$$= \sum_{\alpha,\beta=1}^{m} \Theta^F_{\alpha\beta}(v,\overline{w})\sigma_\alpha \overline{s}_\beta + \sum_{\alpha,\beta=1}^{m} \sum_{\varepsilon=m+1}^{r} a_{\alpha\varepsilon}(v)\overline{a_{\beta\varepsilon}(w)}\sigma_\alpha \overline{s}_\beta.$$

This implies (1).

Similarly, if $\nu, \mu = m+1, ..., r$,

$$\Theta^E_{\nu\mu} = \Theta^Q_{\nu\mu} + \sum_{\alpha=1}^{m} \overline{a}_{\alpha\nu} \wedge a_{\alpha\mu} = \Theta^Q_{\nu\mu} - \sum_{\alpha=1}^{m} a_{\alpha\mu} \wedge \overline{a}_{\alpha\nu}.$$

This implies (2) since if $\varphi = \sum_{\nu=m+1}^{r} \varphi_\nu e_\nu$, $\psi = \sum_{\nu=m+1}^{r} \psi_\nu e_\nu$, then

$$h((A_w)^*(\varphi), (A_v)^*(\psi)) = \sum_{\alpha=1}^{m} h((A_w)^*(\varphi), e_\alpha)\overline{h((A_v)^*(\psi), e_\alpha)}$$
$$= \sum_{\alpha=1}^{m} h(\varphi, A_w(e_\alpha))\overline{h(\psi, A_v(e_\alpha))} = \sum_{\nu,\mu=m+1}^{r} \sum_{\alpha=1}^{s} \varphi_\nu \overline{a_{\alpha\nu}(w)}\, \overline{\psi}_\mu a_{\alpha\mu}(v).$$

□

Identity (2.11) implies that if $A = 0$, then $D^E|A^0(Q) = D^Q$. In this case every holomorphic section of Q is a holomorphic section of E. Hence Q is a holomorphic subbundle of E. Conversely, if Q is a holomorphic subbundle, $A = 0$ as we have noted above. Therefore $E \cong F \oplus Q$ as holomorphic bundles if and only if $A = 0$.

Corollary 2.8.1 *If* $v \in T_x^{1,0}M$, $\sigma \in F_x$, $\varphi \in Q_x$,

(a) $K^F(v; \sigma) \leq K^E(v; \sigma)$

(b) $K^Q(v; \varphi) \geq K^E(v; \varphi)$.

By Proposition 2.8.1, we have equality in inequality (a) exactly when $A_v(\sigma) = 0$. In particular, the equality in (a) holds for every v and every σ if and only if $A = 0$. Similarly for (b).

Corollary 2.8.1 shows that the curvature of the canonical connection decreases when passing to a submanifold and increases when passing to a quotient bundle. This property is an essential difference of holomorphic bundles from smooth ones (the curvature of the unit sphere is bigger than the curvature of the ambient Euclidean space).

Example 2.8.1 Let M be a complex submanifold of $\mathbb{C}^n$ endowed with the Hermitian metric induced by the standard metric of $\mathbb{C}^n$. Then T^hM is a holomorphic subbundle of $T^h\mathbb{C}^n$, hence $K^M(v; w) \leq K^{\mathbb{C}^n}(v; w) = 0$ for every $v \in T_x^{1,0}M$ and $w \in T_x^hM \cong T_x^{1,0}M$, $x \in M$.

As we have seen in the preceding paragraph, if M is a Riemann surface, then $R^M(v, \overline{v})w = \frac{1}{2}K|v|^2w$ where K is the Gauss curvature and $|v|$ is the norm with respect to the metric of M. Hence $K^M(v; w) = \frac{1}{2}K|v|^2|w|^2$. In particular, if M is a one-dimensional complex submanifold of $\mathbb{C}^n$, then $K \leq 0$ since $K^M(v; w) \leq 0$.

Let us stress again that an analogous statement is not true in the real case, the unite sphere in $\mathbb{R}^3$ is of Gauss curvature $K = 1$.

Example 2.8.2 Let $E \to M$ be a holomorphic vector bundle of rank r. Suppose that there exist global holomorphic sections $\sigma_1, \ldots, \sigma_k$ of E such that for every point $x \in M$ the fibre E_x is generated by the vectors $\sigma_1(x), \ldots, \sigma_k(x)$. Then we can define a surjective morphism $f : \underline{\mathbb{C}}^k = M \times \mathbb{C}^k \to E$ by $f(x, \lambda) = \sum_{i=1}^{k} \lambda_i \sigma_i(x)$,

$\lambda = (\lambda_1, ..., \lambda_k) \in \mathbb{C}^k$. Clearly, $rank\ f|\underline{\mathbb{C}}_x^k = dim\ f(\underline{\mathbb{C}}_x^k) = dim\ E_x = r = const.$ Then, by Chap. 1, Proposition 1.7.4, $Ker\ f$ can be endowed with the structure of a holomorphic subbundle of the trivial bundle $\underline{\mathbb{C}}^k$. Therefore the bundle E is isomorphic to the quotient bundle $\underline{\mathbb{C}}^k/Ker\ f$. Let, as above, Q be the orthogonal complement of $Ker\ f$ with respect to the standard metric of $\underline{\mathbb{C}}^k$ endowed with the holomorphic structure of $\underline{\mathbb{C}}^k/Ker\ f$. The holomorphic bundles Q and E are isomorphic, and we can transfer the metric h on Q to E. Then, by Corollary 2.8.1, $K^E = K^Q \geq K^{\underline{\mathbb{C}}^k}|Q = 0$.

Thus, every holomorphic bundle that admits a finite number of global sections generating every fibre admits a Hermitian metric of non-negative curvature.

The connection between the sign of the curvature and the existence of global sections is quite essential in the holomorphic vector bundle theory.

2.9 Existence of Special Frames for a Connection

The next statements can be used to simplify computations involving covariant differentiation.

Lemma 2.9.1 *Let $\pi : E \to M$ be a smooth $\mathbb{K}$-vector bundle and D a connection on E. For every $a \in E_p$, $p \in M$, there exists a (global) section s of E such that*

$$s(p) = a \text{ and } Ds|_p = 0.$$

Proof Let $e_1, ..., e_r$ be a frame of E in a neighbourhood of a point p, and let $s = \sum_{i=1}^r \lambda_i e_i$ be a section of E. Then

$$Ds = \sum_{j=1}^{r}[d\lambda_j + \sum_{i=1}^{r}\lambda_i\theta_{ij}] \otimes e_j,$$

where $[\theta_{ij}]$ is the matrix of the connection D. Let $x_1, ..., x_n$ be a coordinate system of M in a neighbourhood of p such that $x_1(p) = \cdots = x_n(p) = 0$. Let $\theta_{ij} = \sum_{k=1}^n \Gamma_{ij}^k dx_k$. If $a = \sum_{i=1}^r a_i e_i$, the conditions $s(p) = a$, $Ds|_p = 0$ take the form

$$\lambda_j(p) = a_j, \quad \frac{\partial \lambda_j}{\partial x_k}(p) + \sum_{i=1}^{r}\lambda_i(p)\Gamma_{ij}^k(p) = 0, \quad j = 1, ..., r,\ k = 1, ..., n.$$

An obvious solution of these equations for λ_j is

$$\lambda_j = -\sum_{k=1}^{n}(\sum_{i=1}^{r} a_i\Gamma_{ij}^k(p))x_k + a_j.$$

In this way, we can construct a local section s of E that has the desired properties. As we have seen in Sect. 2.1, s can be extended to a globally defined section of E coinciding with s in a neighbourhood of p. It is clear that the extension also has the desired properties. □

Lemma 2.9.2 *Let $\pi : E \to M$ be a holomorphic vector bundle and D a connection on E consistent with the complex structure. Then for every $a \in E_p$, $p \in M$, there exists a (local) holomorphic section s of E defined in a neighbourhood of the point p such that*

$$s(p) = a \text{ and } Ds|_p = 0.$$

Proof Let $e_1, \ldots, e_r$ be a holomorphic frame of E in a neighbourhood of p, and $z_1, \ldots, z_n$ a (complex) coordinate system of M around p such that $z_1(p) = \cdots = z_n(p) = 0$. The connection matrix of D with respect to the frame $e_1, \ldots, e_r$ is of type $(1, 0)$, hence it is of the form $\theta_{ij} = \sum_{k=1}^{n} \Gamma_{ij}^{k} dz_k$. Now, a reasoning similar to that in the proof of the preceding lemma gives the desired result. □

Remark 2.9.1 In contrast to the smooth case, a local holomorphic section s may not be extended to a global holomorphic section. The procedure in Sect. 2.1 for extending smooth sections cannot be applied since there is no holomorphic function φ with compact support on a complex manifold M equal to 1 in a neighbourhood of a point $p \in M$. Any such function is zero on an open non-empty subset of the connected component of M containing p, hence it is zero on this component by the unique holomorphic extension theorem.

Corollary 2.9.1 *Let $\pi : E \to M$ be a smooth vector bundle, D a connection on E, and $a_1, \ldots, a_r$ a basis of a fibre E_p, $p \in M$. Then there exists a frame $s_1, \ldots, s_r$ of E in a neighbourhood of p such that*

$$s_i(p) = a_i \text{ and } Ds_i|_p = 0, \quad i = 1, \ldots, r.$$

A similar statement is true in the case of a holomorphic vector bundle provided the connection is consistent with the complex structure of E.

Proof By Lemmas 2.9.1 and 2.9.2, there exist local sections $s_1, \ldots, s_r$ of E with the properties $s_i(p) = a_i$ and $Ds_i|_p = 0$. The values $\{s_i(p)\}$ of these sections are linearly independent, hence there is a neighbourhood U of p such that the vectors $\{s_i(q)\}$ are linearly independent at every point $q \in U$. □

Now, we give an example of how one can use Lemma 2.9.1 (or Lemma 2.9.2) and Corollary 2.9.1 in order to simplify computations.

Proposition 2.9.1 *Let ∇ be a connection with zero torsion on a manifold M. Let ∇^E be a connection on a vector bundle $E \to M$. Denote by D the connection on the bundle $Hom(TM \otimes TM, E)$ induced by the connections ∇ and ∇^E. Then, considering the curvature tensor R of ∇^E as a section of this bundle, if $X, Y, Z \in T_pM$ and $s \in E_p$, $p \in M$,*

$$(D_X R)(Y, Z)s + (D_Y R)(Z, X)s + (D_Z R)(X, Y)s = 0.$$

Proof By Corollary 2.9.1, for every point $p \in M$, there exists a frame $e_1, ..., e_r$ defined in a neighbourhood of p such that $\nabla^E_U e_i = 0$ for every $U \in T_pM$ and $i = 1, ..., r$. Let $\theta = [\theta_{ij}]$ and $\Theta = [\Theta_{ij}]$ be the connection and the curvature matrices of ∇^E with respect to the frame $e_1, ..., e_r$. Then, at the point p, $\theta = 0$, hence $d\Theta = [\theta, \Theta] = 0$ by Problem 2.7.5. According to Lemma 2.9.1, we can extend the tangent vectors X, Y, Z to vector fields in a neighbourhood of p, denoted by the same symbols, such that $\nabla X|_p = \nabla Y|_p = \nabla Z|_p = 0$. Then, at the point p,

$$\begin{aligned}(D_X R)(Y, Z)e_i &= \nabla^E_X R(Y, Z)e_i - R(\nabla_X Y, Z)e_i - R(Y, \nabla_X Z)e_i - R(Y, Z)\nabla^E_X e_i \\ &= \nabla^E_X R(Y, Z)e_i.\end{aligned}$$

By Sect. 2.7, identity (2.10), $R(Y, Z)e_i = \sum_{j=1}^{e} \Theta_{ij}(Y, Z)e_j$. Hence, at p,

$$\begin{aligned}\nabla^E_X R(Y, Z)e_i &= \sum_{j=1}^{r} (X(\Theta_{ij}(Y, Z))e_j + \Omega_{ij}(Y, Z)\nabla^E_X e_j) = \sum_{j=1}^{r} X(\Theta_{ij}(Y, Z))e_j \\ &= \sum_{j=1}^{r} (\nabla_X \Theta_{ij})(Y, Z))e_j.\end{aligned}$$

Then

$$\begin{aligned}&(D_X R)(Y, Z)e_i + (D_Y R)(Z, X)e_i + (D_Z R)(X, Y)e_i \\ &= \sum_{j=1}^{r} [(\nabla_X \Theta_{ij})(Y, Z)) + (\nabla_Y \Theta_{ij})(Z, X) + (\nabla_Z \Theta_{ij})(X, Y)))e_j \\ &= \sum_{j=1}^{r} (d\Theta_{ij})(X, Y, Z)e_j = 0.\end{aligned}$$

This implies the proposition. □

Proposition 2.9.2 *Let $\pi : E \to M$ be a smooth vector bundle equipped with a Euclidean or Hermitian metric. Let $a_1, ..., a_r$ be an orthonormal basis of a fibre E_p, $p \in M$. If D is a metric connection on E, there exists an orthonormal frame $s_1, ..., s_r$ of E such that*

$$s_i(p) = a_i \text{ and } Ds_i|_p = 0, \quad i = 1, ..., r.$$

Proof Let $\sigma_1, ..., \sigma_r$ be a frame with $\sigma_i(p) = a_i$, $D\sigma_i|_p = 0$, $i = 1, ..., r$. Then the frame $s_1, \ldots, s_r$ obtained from $\sigma_1, \ldots, \sigma_r$ by the Gram-Schmidt orthogonalization possesses the desired properties. Indeed, let g be the metric of E. Set $\widetilde{s}_1 = \sigma_1$ and $s_1 = g(\widetilde{s}_1, \widetilde{s}_1)^{-\frac{1}{2}}\widetilde{s}_1$. For every $X \in T_pM$, we have $Xg(\widetilde{s}_1, \widetilde{s}_1) = g(D_X\sigma_1, \sigma_1)_p + g(\sigma_1, D_X\sigma_1)_p = 0$, hence $X(g(\widetilde{s}_1, \widetilde{s}_1)^{-\frac{1}{2}}) = 0$. Then, using the Leibniz rule for connections and the fact that $D\widetilde{s}_1|_p = 0$, we see that $Ds_1|_p = 0$. Further, let $\widetilde{s}_2 = \sigma_2 + \alpha_{21}s_1$, where $\alpha_{21} = -g(\sigma_2, s_1)$. Then $X(\alpha_{21}) = -g(D_X\sigma_2, s_1)_p - g(\sigma_2, D_Xs_1)_p = 0$. Therefore $D_X\widetilde{s}_2 = D_X\sigma_2 + X(\alpha_{21})D_Xs_1 = 0$. Hence if $s_2 = g(\widetilde{s}_2, \widetilde{s}_2)^{-\frac{1}{2}}\widetilde{s}_2$, then $D_Xs_2 = 0$. We see in the same way that $D_Xs_3 = \cdots = D_Xs_r = 0$. Moreover, it is clear that $s_i(p) = \sigma_i(p)$, hence $s_i(p) = a_i$. □

Remark 2.9.2 If E is a Hermitian holomorphic vector bundle there may not be a holomorphic orthonormal frame $s_1, ..., s_r$. For example, if D is the canonical connection of E and $Z \in T^{1,0}M$, for any such a frame $D_{\overline{Z}}s_i = 0$ and $0 = Z(\delta_{ij}) = g(D_Zs_i, s_j) + g(s_i, D_{\overline{Z}}s_j) = g(D_Zs_i, s_j)$, $i, j = 1, ..., r$. The latter identity implies $D_Zs_i = 0$, hence $Ds_i = 0$, $i = 1, ..., r$. It follows that $D^2s_i = 0$, i.e., the curvature tensor of R is zero. But this condition is not always satisfied: if E is the holomorphic tangent bundle of S^2, the curvature K is positive, thus $R \neq 0$.

As we shall see, if the curvature tensor of a connection D on a smooth vector bundle $E \to M$ vanishes, then there exists a frame $e_1, ..., e_r$ around every point of M such that $De_i = 0$, $i = 1, ..., r$ (Proposition 2.11.4). If, in addition, a metric on E is given and the connection is compatible with this metric, then there exists an orthonormal frame with this property; it is enough to apply the Gram-Schmidt orthogonalization to the frame $e_1, ..., e_r$. If D is the canonical connection of a Hermitian holomorphic bundle, the identity $De_i = 0$ implies clearly $D''e_i = 0$, hence the sections e_i are holomorphic.

The following statement gives an idea of local nature showing how the curvature of the canonical connection is an obstruction to the existence of an orthonormal holomorphic frame.

Proposition 2.9.3 *Let $E \to M$ be a Hermitian holomorphic vector bundle with metric h and curvature tensor of the canonical connection R. Let $a_1, ..., a_r$ be an orthonormal basis of a fibre E_p, $p \in M$. Then for every coordinate system $z = (z_1, ..., z_n)$ of M around the point p with $z(p) = 0$, there exists a holomorphic frame $e_1, ..., e_r$ in a neighbourhood of p such that*

(1) $e_k(p) = a_k$ and $De_k|_p = 0, \quad k = 1, ..., r$,

(2) $h(e_k, e_l) = \delta_{kl} - \sum_{\alpha,\beta=1}^{n} c_{\alpha\beta kl}z_\alpha\overline{z}_\beta + o(|z|^2)$,

$$\text{where } c_{\alpha\beta kl} = h(R(\frac{\partial}{\partial z_\alpha}, \frac{\partial}{\partial \overline{z}_\beta})e_k, e_l)|_p.$$

Proof By Corollary 2.9.1, there exists a holomorphic frame $s_1, ..., s_r$ in a neighbourhood of the point p such that $s_k(p) = a_k$ and $Ds_k|_p = 0$, $k = 1, ..., r$. Expand the function $h(s_k, s_l)$ in Taylor series around the point p. Note that $h(s_k, s_l)(p) = h(a_k, a_l) = \delta_{kl}$. Furthermore, for every $X \in T_pM$, $Xh(s_k, s_l) = h(D_Xs_k, s_l)_p +$

$h(s_l, Ds_k)_p = 0$. Thus, the first derivatives of $h(s_k, s_l)$ at p with respect to the local coordinates are zero. Therefore the Taylor expansion of the function $h(s_k, s_l)$ around p has the form

$$h(s_k, s_l) = \delta_{kl} + o(|z|) = \delta_{kl} + \sum_{\alpha,\beta=1}^{n} (c_{\alpha\beta kl} z_\alpha \overline{z}_\beta + c'_{\alpha\beta kl} z_\alpha z_\beta + c''_{\alpha\beta kl} \overline{z}_\alpha \overline{z}_\beta) + o(|z|^2),$$

where $c_{\alpha\beta kl}, c'_{\alpha\beta kl}, c''_{\alpha\beta kl} \in \mathbb{C}$. We have $c_{\alpha\beta kl} = \overline{c_{\beta\alpha lk}}$ and $c'_{\alpha\beta kl} = \overline{c''_{\alpha\beta lk}}$ since $h(s_k, s_l) = \overline{h(s_l, s_k)}$. Set

$$e_k = s_k - \sum_{l=1}^{r} \sum_{\alpha,\beta} c'_{\alpha\beta kl} z_\alpha z_\beta s_l, \quad k = 1, ..., r.$$

It is obvious that the sections $e_1, ..., e_r$ are holomorphic. Since $e_k(p) = s_k(p) = a_k$, $e_1, ..., e_r$ is a frame of E in a neighbourhood of the point p. Taking into account the identity $c''_{\alpha\beta lk} = \overline{c'_{\alpha\beta kl}}$, we find

$$h(e_k, e_l) = \delta_{kl} + \sum_{\alpha,\beta=1}^{n} c_{\alpha\beta kl} z_\alpha \overline{z}_\beta + o(|z|^2). \tag{2.12}$$

Set for brevity $h_{kl} = h(e_k, e_l)$ and $h = [h_{kl}]$. Let θ be the matrix of the canonical connection with respect to the frame $e_1, ..., e_r$, and let Θ be the curvature matrix. Then $\theta = \partial h \cdot h^{-1}$ and $\Theta = \overline{\partial}\theta = \overline{\partial}\partial h \cdot h^{-1} - \partial h \wedge h^{-1} \cdot \overline{\partial} h \cdot h^{-1}$. From the identity (2.12), we find

$$\partial h_{kl} = \sum_{\alpha,\beta=1}^{n} c_{\alpha\beta kl} \overline{z}_\beta dz_\alpha + o(|z|),$$
$$\overline{\partial}\partial h_{kl} = \sum_{\alpha,\beta=1}^{n} c_{\alpha\beta kl} d\overline{z}_\beta \wedge dz_\alpha + o(1).$$

We have $\partial h_{kl}|_p = 0$ since $z_\beta(p) = 0$, $\beta = 1, ..., n$. It follows $\theta|_p = 0$, hence $De_k|_p = 0$, $k = 1, ..., r$. Moreover

$$\Theta_{kl}|_p = \overline{\partial}\partial h_{kl}|_p = \sum_{\alpha,\beta=1}^{n} c_{\alpha\beta kl} d\overline{z}_\beta \wedge dz_\alpha|_p.$$

Thus,

$$h(R(\frac{\partial}{\partial z_\alpha}, \frac{\partial}{\partial \overline{z}_\beta})e_k, e_l)|_p = \Theta_{kl}(\frac{\partial}{\partial z_\alpha}, \frac{\partial}{\partial \overline{z}_\beta})|_p = -c_{\alpha\beta kl}.$$

□

Remark 2.9.3 If the frame $e_1, ..., e_r$ is orthonormal, identity (2) of Proposition 2.9.3 implies $R_p = 0$. We may say that the curvature is an obstruction of second order for orthonormality of the frame $e_1, ..., e_r$; by (2), obstruction of first order does not exist.

Corollary 2.9.2 *Let M be a Hermitian complex manifold with metric h and curvature tensor of its canonical connection R. Let $a_1, ..., a_r$ be an orthonormal basis of a tangent space $T_p^h M$, $p \in M$. Then there exists a coordinate system $w = (w_1, ..., w_n)$ of M in a neighbourhood of such that $w(p) = 0$ and*

$$(1)\ \frac{\partial}{\partial w_k}(p) = a_k \text{ and } D\frac{\partial}{\partial w_k}|_p = 0, \quad k = 1, ..., n$$

$$(2)\ h(\frac{\partial}{\partial w_k}, \frac{\partial}{\partial w_l}) = \delta_{kl} - \sum\nolimits_{\nu,\mu=1}^{n} c_{\nu\mu kl} w_\nu \overline{w}_\mu + o(|w|^2),$$

$$\text{where } c_{\nu\mu kl} = h(R(\frac{\partial}{\partial w_\nu}, \frac{\partial}{\partial \overline{w}_\mu})\frac{\partial}{\partial w_k}, \frac{\partial}{\partial w_l})|_p.$$

Proof Let $z = (z_1, ..., z_n)$ be a coordinate system of M in a neighbourhood of p with $z(p) = 0$. Let $e_1, ..., e_n$ be a holomorphic frame of $T^h M$ with the properties (1) and (2) of Proposition 2.9.3. The vector fields e_k have the representation $e_k = \sum_{l=1}^n \lambda_{kl} \frac{\partial}{\partial z_l}$, where the functions λ_{kl} are holomorphic and $det\,[\lambda_{kl}] \neq 0$. Let $[\lambda^{kl}]$ be the inverse matrix of the matrix $[\lambda_{kl}]$. By the standard rule for computing an inverse matrix, the functions λ^{kl} are holomorphic. Set $w_j = \sum_{l=1}^n \lambda^{lj} z_l$. The functions w_j are holomorphic and $det \left[\frac{\partial w_j}{\partial z_l}\right] = det\,[\lambda^{lj}] \neq 0$. Therefore $w = (w_1, ..., w_n)$ is a coordinate system of M in a neighbourhood of p. Since $z_l = \sum_{j=1}^n \lambda_{jl} w_j$, we have $\frac{\partial}{\partial w_j} = \sum_{l=1}^n \frac{\partial z_l}{\partial w_j}(p)\frac{\partial}{\partial z_l}(p) = \sum_{l=1}^n \lambda_{jl}\frac{\partial}{\partial z_l}(p) = e_j$. Therefore property (1) holds. It is easy to verify that property (2) also is satisfied. □

2.10 Horizontal Subbundle Determined by a Connection

Let $\pi : E \to M$ be a smooth $\mathbb{K}$-vector bundle. The differential $\pi_* : TE \to TM$ can be considered as a bundle morphism $TE \to \pi^* TM$. It is surjective, hence of constant rank. Therefore $Ker\,\pi_*$ is a vector subbundle of TE (Chap. 1, Proposition 1.7.4). The fibre $Ker\,\pi_{*a}$ of this bundle at a point $a \in E$ is the tangent space of the submanifold $\pi^{-1}(\pi(a))$ = the fibre $E_{\pi(a)}$ of E. The bundle $Ker\,\pi_*$ is called **vertical**, and its elements vertical vectors.

Denote the vertical bundle by $\mathcal{V}$. The fibre $E_{\pi(a)}$ is a vector space, so its tangent space $\mathcal{V}_a = T_a E_{\pi(a)}$ can canonically be identified with $E_{\pi(a)}$.

In general, there is no canonical procedure to choose a subbundle that complements $\mathcal{V}$ to the tangent bundle TE. However, if a connection D on E is given such a procedure exists.

Let $a \in E$, $p = \pi(a)$ and let s be a section of E with $s(p) = a$ and $Ds|_p = 0$. Then $\pi_* \circ s_* = Id$ since $\pi \circ s = Id$, hence the space $s_*(T_pM) \subset T_aE$ is of dimension $n = dim\, M$ and does not contain non-zero vertical vectors. This space does not depend on the choice of the section s. To show this, we compute $s_*(X)$, $X \in T_pM$, in local coordinates of the manifold E defined by means of a local trivialization of the bundle E. Let $f = (s_1, ..., s_r)$ be a frame of E defined in a coordinate neighbourhood U of the point p, and let $x = (x_1, ..., x_n)$ be a local coordinate system of M on U. For $e \in \pi^{-1}(U)$, set $e = \sum_{i=1}^r y_i(e)s_i \circ \pi(e)$. Then $\widetilde{x}_1 = x_1 \circ \pi, ..., \widetilde{x}_n = x_n \circ \pi, y_1, ..., y_r$ is a local coordinate system of E on $\pi^{-1}(U)$ (it is the composition of the diffeomorphism $\pi^{-1}(U) \cong U \times \mathbb{K}^r$ determined by the frame f and the standard basis of $\mathbb{K}^r$ with the coordinate system $x \times Id$ on $U \times \mathbb{K}^r$).

For $a \in E$, denote the standard isomorphism $\mathcal{V}_a = T_aE_{\pi(a)} \cong E_{\pi(a)}$ by χ_a. Let $\chi : \mathcal{V} \to \pi^*E = \{(a, e) \in E \times E : e \in E_{\pi(a)}\}$ be the map defined by $\chi(v) = (a, \chi_a(v))$ for $v \in \mathcal{V}_a$. This map sends the frame $\dfrac{\partial}{\partial y_j}$, $j = 1, ..., r$, of $\mathcal{V}$ to the frame $(Id, s_j \circ \pi)$ of π^*E. Hence χ is a smooth map that is a linear isomorphism on the fibres. Thus, χ is an isomorphism of bundles (Chap. 1, Proposition 1.2.1)

Let $a = \sum_{i=1}^r a_i s_i(p)$ and $s = \sum_{i=1}^r \lambda_i s_i$; here $a_i = y_i(a)$ and $\lambda_i = y_i \circ s$. Let $[\theta_{ij}]$ be the matrix of the connection D with respect to the frame $s_1, ..., s_r$. Set $\theta_{ij} = \sum_{k=1}^n \Gamma_{ij}^k dx_k$. As we have seen in the proof of Lemma 2.9.1, identities $s(p) = a$, $Ds|_p = 0$ imply

$$\frac{\partial \lambda_j}{\partial x_k}(p) + \sum_{i=1}^r a_i \Gamma_{ij}^k(p) = 0, \quad j = 1, ..., r,\ k = 1, ..., n.$$

Therefore

$$\begin{aligned} s_*(\frac{\partial}{\partial x_k}(p)) &= \sum_{l=1}^n \frac{\partial(\widetilde{x}_l \circ s)}{\partial x_k}(p)\frac{\partial}{\partial \widetilde{x}_l}(a) + \sum_{j=1}^r \frac{\partial(y_j \circ s)}{\partial x_k}(p)\frac{\partial}{\partial y_j}(a) \\ &= \frac{\partial}{\partial \widetilde{x}_k}(a) - \sum_{i,j=1}^r a_i \Gamma_{ij}^k(p)\frac{\partial}{\partial y_j}(a). \end{aligned} \tag{2.13}$$

This shows that the vectors $s_*(\dfrac{\partial}{\partial x_k}(p))$ do not depend on the choice of s. It follows from (2.13) that these vectors are linearly independent. Hence $\mathcal{H}_a = s_*(T_pM)$ is a n-dimensional subspace of T_aE that does not intersect $\mathcal{V}_a$. Since $dim\, \mathcal{V}_a + dim\, \mathcal{H}_a = dim\, T_aE$, we have $T_aE = \mathcal{V}_a \oplus \mathcal{H}_a$. The space $\mathcal{H}_a$ is called **horizontal**. Since $\mathcal{V}_a = Ker\, \pi_{*a}$, the map $\pi_*|\mathcal{H}_a$ is an isomorphism of the vector space $\mathcal{H}_a$ onto the space $T_{\pi(a)}M$.

Set

$$H_k = \frac{\partial}{\partial \widetilde{x}_k} - \sum_{i,j=1}^r y_i(\Gamma_{ij}^k \circ \pi)\frac{\partial}{\partial y_j}, \quad k = 1, ..., n. \tag{2.14}$$

At each point $a \in \pi^{-1}(U)$, these smooth vector fields form a basis of the horizontal space $\mathcal{H}_a$. Therefore $\mathcal{H} = \dot{\cup}\,\{\mathcal{H}_a : a \in E\}$ is a subbundle of TE.

If $X \in T_pM$, the horizontal vector $(\pi_*|\mathcal{H}_a)^{-1}(X)$ is called the **horizontal lift** of X and is often denoted by X^h_a. We have $X^h_a = s_*(X)$ since $s_*(X) \in \mathcal{H}_a$ and $\pi_*(s_*(X)) = X$. Identities $\Gamma^k_{ij} = \theta_{ij}(\frac{\partial}{\partial x_k})$ and (2.13) imply

$$X^h_a = \sum_{k=1}^{n} X(x_k)\frac{\partial}{\partial \widetilde{x}_k}(a) - \sum_{i,j=1}^{r} y_i(a)\theta_{ij}(X)\frac{\partial}{\partial y_j}(a). \tag{2.15}$$

Recall that the identity $X^h_a = s_*(X)$ holds provided s is a section of E with the properties $s(p) = a$ and $Ds|_p = 0$. Now, let σ be an arbitrary smooth section of E with $\sigma(p) = a$. Then, taking into account identity (2.15) and the fact that $X(\widetilde{x}_k \circ \sigma) = X(x_k)$, we obtain

$$\begin{aligned}\sigma_*(X) &= \sum_{k=1}^{n} X(\widetilde{x}_k \circ \sigma)\frac{\partial}{\partial \widetilde{x}_k}(a) + \sum_{j=1}^{r} X(y_j \circ \sigma)\frac{\partial}{\partial y_j}(a)\\ &= X^h_a + \sum_{i,j=1}^{r}\left[y_i(a)\theta_{ij}(X)\frac{\partial}{\partial y_j}(a) + \sum_{j=1}^{r} X(y_j \circ \sigma)\frac{\partial}{\partial y_j}(a)\right].\end{aligned} \tag{2.16}$$

Under the canonical isomorphism $\mathcal{V}_a = T_aE_p \cong E_p$, $p = \pi(a)$, the expression into the parentheses goes to the vector

$$\sum_{i,j=1}^{r} y_i(a)\theta_{ij}(X)s_j(p) + \sum_{j=1}^{r} X(y_j \circ \sigma)s_j(p).$$

In fact, this is the vector $D_X\sigma$ since $\sigma = \sum_{i=1}^{r}(y_i \circ \sigma)s_i$. Thus,

$$\sigma_*(X) = X^h_a + D_X\sigma. \tag{2.17}$$

Remark 2.10.1 If E is the tangent bundle of a manifold M, then, in addition to the horizontal lift of any vector $X \in T_pM$, we have a vertical lift of X at the points a with $\pi(a) = p$. This is the image of X under the canonical isomorphism $T_pM \cong T_a(T_pM) = \mathcal{V}_a$. In the general case, there is no vertical lift of a vector in T_pM. But every section σ of a vector bundle E determines a vertical vector field on E defined by $a \to \sigma \circ \pi(a) \in E_{\pi(a)} \cong T_aE_{\pi(a)} = \mathcal{V}_a$.

The horizontal spaces possess also the following property. For $\alpha \in \mathbb{K}$, let $\widehat{\alpha} : E \to E$ be the map defined by $\widehat{\alpha}(v) = \alpha v$. Clearly, $\pi \circ \widehat{\alpha} = \pi$, hence $\pi_* \circ \widehat{\alpha}_* = \pi_*$. Therefore $\widehat{\alpha}_{*a}$ maps $\mathcal{V}_a$ into $\mathcal{V}_{\alpha a}$ whereby

$$\widehat{\alpha}_{*a}(\frac{\partial}{\partial y_j}(a)) = \alpha\frac{\partial}{\partial y_j}(\alpha a), \quad j = 1, \ldots, r. \tag{2.18}$$

Hence if $\alpha \neq 0$, the map $\widehat{\alpha}_{*a}$ is an isomorphism of $\mathcal{V}_a$ onto $\mathcal{V}_{\alpha a}$. Furthermore,

$$\widehat{\alpha}_{*a}(\frac{\partial}{\partial \widetilde{x}_k}(a)) = \frac{\partial}{\partial \widetilde{x}_k}(\alpha a), \quad k = 1, ..., n. \tag{2.19}$$

Then, by (2.14),

$$\widehat{\alpha}_{*a}(H_k(a)) = H_k(\alpha a), \quad k = 1, ..., n.$$

Therefore $\widehat{\alpha}_{*a}$ is an isomorphism of $\mathcal{H}_a$ onto $\mathcal{H}_{\alpha a}$.

Thus, every connection D on a bundle E determines a subbundle $\mathcal{H} = \mathcal{H}^D$ of TE such that

(*i*) $TE = \mathcal{V} \oplus \mathcal{H}$;
(*ii*) $\widehat{\alpha}_{*a}(\mathcal{H}_a) = \mathcal{H}_{\alpha a}$ for every $\alpha \in \mathbb{K}$ and $a \in E$.

These two properties uniquely determine the connection D. More precisely, the following statement holds.

Proposition 2.10.1 *Let $\mathcal{H}$ be a subbundle of the tangent bundle TE of the total space of a vector bundle $\pi : E \to M$ possessing the properties (i) and (ii) above. Then there is a unique connection D on E for which $\mathcal{H}$ is the horizontal bundle of D.*

Proof Let $p_{\mathcal{V}} : TE = \mathcal{V} \oplus \mathcal{H} \to \mathcal{V}$ be the natural projection. For $a \in E$, let $\chi_a : \mathcal{V}_a \to E_{\pi(a)}$ be the canonical isomorphism.

Uniqueness. If D is a connection with horizontal space $\mathcal{H}$, then, by (2.17), for every smooth section s of E and every tangent vector $X \in T_pM$

$$D_X s = \chi_{s(p)} \circ p_{\mathcal{V}_{s(p)}} \circ s_{*p}(X). \tag{2.20}$$

The right-hand side of this identity shows that $D_X s$ is uniquely determined by $\mathcal{H}$.

Existence. If s is a smooth section of E and $X \in T_pM$, define $D_X s$ by identity (2.20). Let $\chi : \mathcal{V} \cong \pi^*E = \{(a, e) \in E \times E : e \in E_{\pi(a)}\}$ be the isomorphism defined by means of the maps χ_a. Denote the projection of $E \times E$ onto the second factor by p_2. Then $Ds = (p_2|\pi^*E) \circ \chi \circ p_{\mathcal{V}} \circ s_*$. Since π^*E is a submanifold of $E \times E$, the map $p_2|\pi^*E$ is smooth and the identity $Ds = (p_2|\pi^*E) \circ \chi \circ p_{\mathcal{V}} \circ s_*$ implies that Ds is a smooth map of TM into E. It is obvious that $D_X s$ is $\mathbb{K}$-linear with respect to X. Hence Ds is a 1-form on M with values in E, i.e., D maps $A^0(E)$ into $A^1(E)$. In order to check whether Leibniz's rule holds for D, take a smooth function f on M and set $\sigma = fs$. Also, let $p \in M$ and set $\alpha = f(p)$ and $a = s(p)$. We have $\chi_a(\frac{\partial}{\partial y_j}(a)) = s_j \circ \pi(a)$, $j = 1, ..., r$, which together with identity (2.18) implies

$$\chi_{\alpha a} \circ \widehat{\alpha}_{*a}|\mathcal{V}_a = \widehat{\alpha} \circ \chi_a. \tag{2.21}$$

Moreover, by (ii), $\widehat{\alpha}_{*a}$ maps $\mathcal{H}_a$ in $\mathcal{H}_{\alpha a}$, hence

$$p_{\mathcal{V}_{\alpha a}} \circ \widehat{\alpha}_{*a} = \widehat{\alpha}_{*a} \circ p_{\mathcal{V}_a}. \tag{2.22}$$

We have $\widetilde{x}_k \circ \sigma = \widetilde{x}_k \circ s$ and $y_j \circ \sigma = f(y_j \circ s)$. Then the first identity in (2.16) and identities (2.18), and (2.19) imply

$$\sigma_{*p}(X) = \widehat{\alpha}_{*a} \circ s_{*p}(X) + X(f) \sum_{j=1}^{r} y_j(a) \frac{\partial}{\partial y_j}(\alpha a).$$

Hence

$$D_X \sigma = \chi_{\alpha a} \circ p_{\mathcal{V}_{\alpha a}} \circ \widehat{\alpha}_{*a} \circ s_{*p}(X) + X(f) \sum_{j=1}^{r} y_j(a) s_j(p).$$

By (2.22) and (2.21)

$$\chi_{\alpha a} \circ p_{\mathcal{V}_{\alpha a}} \circ \widehat{\alpha}_{*a} = \chi_{\alpha a} \circ \widehat{\alpha}_{*a} \circ p_{\mathcal{V}_a} = \widehat{\alpha} \circ \chi_a \circ p_{\mathcal{V}_a}.$$

Therefore

$$D_X \sigma = \alpha \chi_a \circ p_{\mathcal{V}_a} \circ s_{*p}(X) + X(f) a = f(p) D_X s + X(f) s(p).$$

This proves that D is a connection on E.

Denote for the moment the horizontal space of D at a point $a \in E$ by $\mathcal{H}_a^D$. Let $p = \pi(a)$, and take a section s of E with $s(p) = a$, $Ds|_p = 0$. Then identity (2.20) implies that for every $X \in T_pM$, $p_{\mathcal{V}_a}(s_{*p}(X)) = 0$, i.e., $s_{*p}(X) \in \mathcal{H}_a$. This shows that $\mathcal{H}_a^D \subset \mathcal{H}_a$. By property (i), $dim\, \mathcal{H}_a = dim\, E - dim\, \mathcal{V}_a = dim\, M = dim\, \mathcal{H}_a^D$. Therefore $\mathcal{H}_a^D = \mathcal{H}_a$. □

Remark 2.10.2 Condition (i) can be stated in the language of exact sequences of bundles. Recall that a sequence

$$\cdots \longrightarrow A \xrightarrow{f} B \xrightarrow{g} C \longrightarrow \cdots$$

of bundles over the same manifold with morphisms between them is called exact if $Im\, f = Ker\, g$ for every morphisms f, g in the sequence.

Let $\iota : \mathcal{V} \hookrightarrow TE$ be the inclusion map. Then the sequence of bundles over E

$$0 \longrightarrow \mathcal{V} \xrightarrow{\iota} TE \xrightarrow{\pi_*} \pi^*TM \longrightarrow 0$$

is exact, where 0 means the trivial bundle $E \times \{0\}$. The condition for the existence of a subbundle $\mathcal{H}$ such that $TE = \mathcal{V} \oplus \mathcal{H}$ can be expressed as the condition for the existence of a morphism $h : \pi^*TM \to TE$ with the property $\pi_* \circ h = Id$ (if h is given, set $\mathcal{H} = Im\, h$). This version of the notion of a horizontal bundle, i.e., a connection, is germane to generalizations.

Let $\pi : E \to M$ be a vector bundle with a connection D.

Problem 2.10.1 Let $a \in E$ and let X, Y be vector fields in a neighbourhood of the point $\pi(a)$. Show that, if we identify the vertical space $\mathcal{V}_a$ with the fibre $E_{\pi(a)}$, then

$$[X^h, Y^h]_a = [X, Y]^h_a - R(X, Y)a,$$

where R is the curvature tensor of the connection D.

Problem 2.10.2 For any section σ of E, denote by $\widetilde{\sigma}$ the vertical vector field on E defined by $\widetilde{\sigma}(a) = \sigma \circ \pi(a)$, where it is understood that the fibre $E_{\pi(a)}$ is identified with the vertical space $\mathcal{V}_a$. Show that

$$[X^h, \widetilde{\sigma}] = \widetilde{D_X\sigma}.$$

Now, let $\pi : E \to N$ be a vector bundle with a connection D and $f : M \to N$ a smooth map. Denote by $\widetilde{D}$ the connection on the pull-back bundle $\widetilde{\pi} : \widetilde{E} = f^*E$ yielded by D. Let $\widetilde{f} : \widetilde{E} \to E$ be the standard map that is an isomorphism on the fibres. Usually, we identify every section $\widetilde{s}$ of $\widetilde{E}$ with the map $s = \widetilde{f} \circ \widetilde{s} : M \to E$, and write $\widetilde{D}s$ instead of $\widetilde{f}(\widetilde{D}\widetilde{s})$. However, we do not use these identifications in the next statement.

Lemma 2.10.1 *Let $p \in M$ and $X \in T_pM$. If $\widetilde{s}$ is a section of f^*E and $s = \widetilde{f} \circ \widetilde{s}$, then*

$$s_*(X) = \widetilde{f}(\widetilde{D}_X\widetilde{s}) + (f_*X)^h_{s(p)},$$

where $\widetilde{f}(\widetilde{D}_X\widetilde{s}) \in E_{f(p)}$ is considered as a vertical vector of E at the point $s(p)$ and the horizontal lift is taken with respect to the connection D of E.

Proof Let $(U, x_1, ..., x_n)$ be a chart of N at the point $f(p)$ such that a frame $s_1, ..., s_r$ of E with $Ds_j|_{f(p)} = 0$, $j = 1, ..., r$, is defined on U. As above, denote by $\{\widetilde{x}_k = x_k \circ \pi, y_j\}$ the local coordinates of E on $\pi^{-1}(U)$ determined by x_k and s_j, $k = 1, ..., n$, $j = 1, ..., r$. Then $s = \sum_{j=1}^r (y_j \circ s)(s \circ f)$, hence

$$\widetilde{f}(\widetilde{D}_X\widetilde{s}) = \sum_{j=1}^r X(y_j \circ s)(s_j \circ f)(p) + \sum_{j=1}^r (y_j \circ s)(p) D_{f_*X} s_j = \sum_{j=1}^r X(y_j \circ s)(s_j \circ f)(p).$$

Moreover, $f_*(X) = \sum_{k=1}^n X(x_k \circ f)\frac{\partial}{\partial x_k}(f(p))$. Since $Ds_j = 0$ at the point $f(p) = \pi(s(p))$, this implies

$$(f_*X)^h_{s(p)} = \sum_{k=1}^n X(x_k \circ f)\frac{\partial}{\partial \widetilde{x}_k}(s(p)).$$

On the other hand

$$s_*(X) = \sum_{k=1}^{n} X(\widetilde{x}_k \circ s)\frac{\partial}{\partial \widetilde{x}_k}(s(p)) + \sum_{j=1}^{r} X(y_j \circ s)\frac{\partial}{\partial y_j}(s(p)).$$

It only remains to note that $X(\widetilde{x}_k \circ s) = X(x_k \circ f)$ in the first sum and that the canonical isomorphism of $E_{\pi(s(p))}$ onto $\mathcal{V}_{f(p)} = T_{f(p)}E_{\pi(s(p))}$ maps the vector $s_j(f(p))$ to $\frac{\partial}{\partial y_j}(s(p))$. □

Corollary 2.10.1 *Let $\widetilde{\mathcal{V}}$ be the vertical subbundle of $T\widetilde{E}$ and $\widetilde{\mathcal{H}}$ its horizontal subbundle with respect to the connection $\widetilde{D}$. Then, for every $a \in \widetilde{E}$, the differential $\widetilde{f}_{*a}$ isomorphically maps $\widetilde{\mathcal{V}}_a$ onto the vertical space $\mathcal{V}_{\widetilde{f}(a)}$ of E and sends $\widetilde{\mathcal{H}}_a$ into the horizontal space $\mathcal{H}_{\widetilde{f}(a)}$ of E with respect to D.*

The map $\widetilde{f}_{*a}|\widetilde{\mathcal{H}}_a$ is injective exactly when the map $f_{*\widetilde{\pi}(a)}$ is injective.

Proof Since the map $\widetilde{f}$ is an isomorphism on the fibres, its differential is an isomorphism of their tangent spaces, i.e., of the vertical spaces. Let $p = \widetilde{\pi}(a)$ and let $\widetilde{s}$ be a section of $\widetilde{E}$ in a neighbourhood of p such that $\widetilde{s}(p) = a$ and $\widetilde{D}\widetilde{s}|_p = 0$. Then, by Lemma 2.10.1, if $X \in T_pM$ and $\widetilde{X}_a^h$ is its horizontal lift with respect to $\widetilde{D}$, we have $\widetilde{f}_*(\widetilde{X}_a^h) = \widetilde{f}_* \circ \widetilde{s}_*(X) = s_*(X) = (f_*X)_{s(p)}^h$. Therefore $\widetilde{f}_*(\widetilde{\mathcal{H}}_a) \subset \mathcal{H}_{\widetilde{f}(a)}$.

The last part of the claim is an obvious consequence of the identity $\widetilde{f}_*(\widetilde{X}_a^h) = (f_*X)_{s(p)}^h$. □

Corollary 2.10.2 *Let $E \to N$ be a vector bundle with a connection D and M a submanifold of N. Then the horizontal bundle of the subbundle $E|M$ of E with respect to the connection induced by D coincides with $\mathcal{H}|(E|M)$, the restriction of the horizontal bundle $\mathcal{H} \to E$ of E to the submanifold $E|M$ of E.*

2.11 Parallel Transport Along Curves with Respect to a Connection

If M and N are smooth manifolds and A is an arbitrary subset of M, a map $A \to N$ will be called smooth if it can be extended to a smooth map $G \to N$ in an open neighbourhood G of A. When it is irrelevant which continuation we are considering, the continuation will be denoted with the same symbol as the given map of A, and often we will not indicate the domain on which the extension is defined.

If M is a smooth manifold, a smooth curve in M is by definition a smooth map $\gamma : [a, b] \to M$ of a finite and closed interval into M. Let $E \to M$ be a vector bundle over M and γ a smooth curve in M. **A section s of E along the curve γ** is by definition a section of the pull-back bundle γ^*E. In other words, $s : [a, b] \to E$

is a map for which there exists an open interval $J \supset [a, b]$ such that γ and s extend to smooth maps on J and $s(t) \in E_{\gamma(t)}$ for every $t \in J$.

Suppose a connection D is given on the bundle E. Denote the connection induced by D on the vector bundle $\widetilde{E} = \gamma^* E$ by $\widetilde{D}$. A section s of E along γ is called **parallel with respect to** D if $\widetilde{D}s = 0$.

Let s be a section of E along γ. For every point $\tau \in [a, b]$, let U_τ be a neighbourhood of the point $\gamma(\tau)$ in M admitting a frame $s_1, ..., s_r$ of the bundle E. Since the map γ is continuous, there is an open interval J_τ containing τ such that $\gamma(\overline{J_\tau}) \subset U_\tau$. Then the sections $s_1 \circ \gamma, ..., s_r \circ \gamma$ on J_τ constitute a frame of $\widetilde{E}$, and s on the interval J_τ can be written as $s = \sum_{i=1}^r f_i(s_i \circ \gamma)$, where f_i are smooth functions. Denote by $[\theta_{ij}]$ the connection matrix of D with respect to the frame $s_1, ..., s_r$. Let $\dfrac{d}{dt}$ be the standard vector field on the real line. Then

$$\begin{aligned}\widetilde{D}_{\frac{d}{dt}} s &= \sum_{i=1}^r \tfrac{df_i}{dt}(s_i \circ \gamma) + \sum_{i=1}^r f_i (D_{\gamma_*(\frac{d}{dt})} s_i) \circ \gamma \\ &= \sum_{j=1}^r \tfrac{df_j}{dt}(s_i \circ \gamma) + \sum_{j=1}^r \sum_{i=1}^r f_i \theta_{ij}(\gamma_*(\tfrac{d}{dt}))(s_i \circ \gamma).\end{aligned}$$

Therefore the condition for s to be parallel along J_τ is

$$\frac{df_j}{dt} + \sum_{i=1}^r f_i \theta_{ij}(\gamma_*(\frac{d}{dt})) = 0, \quad j = 1, ..., r. \tag{2.23}$$

Let $v \in E_{\gamma(\tau)}$ and $v = \sum_{i=1}^r v_i s_i(\gamma(\tau))$. By the theorem for existence and uniqueness of a solution of a linear ODE, there exists a unique set of smooth functions $f_1, ..., f_r$ on the interval J_τ satisfying Eqs. (2.23) and the initial conditions $f_j(\tau) = v_j$, $j = 1, ..., r$; note that the functions f_j are defined on the whole interval J_τ not just in a neighbourhood of τ since the Eqs. (2.23) are linear, see, e.g., Conlon [4], Chap. 2, Theorem 2.8.4 (proved in Appendix C) and Theorem C.4.1. Hence, for every $v \in E_{\gamma(\tau)}$, there exists a unique parallel section s_v of E along the curve $\gamma|J_\tau$ such that $s_v(\tau) = v$.

The interval $[a, b]$ is compact, thus we can find a fine number of intervals $J_{\tau_1}, ..., J_{\tau_m}$ covering $[a, b]$. By the Lebesgue covering lemma, there is a number $r > 0$ such that for every $t \in [a, b]$ the interval $(t - r, t + r)$ lies in some interval $J_l = J_{\tau_l}$. Let $a = t_0 < t_1 < ... < t_{k-1} < t_k = b$ be a partition of the interval $[a, b]$ such that $t_{i-1} - t_i < \frac{1}{2}r$, $i = 1, ..., k$. Then every closed interval $[t_{i-1}, t_i]$ lies in some interval J_l together with an open interval $\Delta_{i-1} \supset [t_{i-1}, t_i]$. Let $v \in E_{\gamma(a)}$. As discussed above, there exists a unique parallel section s_0 of E along the curve $\gamma|\Delta_0$ such that $s_0(a) = v$. Also, there exists a unique parallel section s_1 of E along the curve $\gamma|\Delta_1$ such that $s_1(t_1) = s_0(t_1)$. By the uniqueness theorem for the solution of an ODE, $s_1 = s_0$ on the interval $\Delta_1 \cap \Delta_0$. Thus, by induction, we can define a parallel section s of E along γ such that $s(a) = v$.

Let s' be a parallel section of E along the curve γ with $s'(a) = v$. By the uniqueness theorem for an ODE, $s' = s$ on the interval Δ_0. Since $s'(t_1) = s(t)$, $s' = s$ on Δ_1, etc. Hence $s' = s$ in a neighbourhood of the interval $[a, b]$.

These considerations prove the following.

Proposition 2.11.1 *Let $E \to M$ be a vector bundle with a connection D, and let $\gamma : [a, b] \to M$ be a smooth curve in M. Then, for every $v \in E_{\gamma(a)}$, there exists a unique parallel section s_v of E along γ with respect to D such that $s_v(a) = v$.*

Let $\lambda : [a', b'] \to [a, b]$ be a linear function: $\lambda(t) = ct + d$. Obviously, it follows from (2.23) that if s is a parallel section of E along a curve γ, $s' = s \circ \lambda$ is a parallel section of E along the curve $\gamma' = \gamma \circ \lambda$. Hence the notion of a parallel section along a curve is invariant under affine changes of the curve parameter.

Next, note that given a smooth curve $\gamma : J \to M$ defined in an open interval J, a point $t_0 \in J$, and a vector $v \in E_{\gamma(t_0)}$, there exists a unique parallel section s of E along γ defined on J such that $s(t_0) = v$. To see this, take a finite open interval $[t_0, b] \subset J$ and consider the parallel section s of E along the curve $\gamma|[t_0, b]$ with $s(t_0) = v$. Let $\{b_i\}$ be an increasing sequence of real numbers $b_i > b$ converging to the right end of the interval J (which can be $+\infty$). Then applying successive extensions of s on the intervals $[b, b_1]$, $[b_1, b_2]$,... we get a parallel section on the subinterval of J to the right of the point b. Similarly, s can be extended to a parallel section to the left of the point t_0.

Suppose that the curve γ depends smoothly on a parameter u in an open subset G of $\mathbb{R}^n$, i.e., $\gamma(t, u)$ is a smooth map of $J \times G$ into M, where J is an open interval containing $[a, b]$. Let $t_0 \in [a, b]$, and let $v : G \to E$ be a smooth map such that $v(u) \in E_{\gamma(t_0,u)}$. For $u \in G$, denote by s_u the parallel section of E along the curve $J \ni t \to \gamma(t, u)$ such that $s_u(t_0) = v(u)$. The theorem for smooth dependence on parameters of solutions of ODE's (see again Conlon, Chap. 2, Theorem 2.8.4) implies that the map $s : J \times G \to E$ defined by $s(t, u) = s_u(t)$ is smooth. For every (t, u), $s(t, u) \in E_{\gamma(t,u)}$, thus s can be considered as a smooth section of the bundle $\gamma^* E$.

Now, let $\gamma : [a, b] \to M$ be a smooth curve, and let $T_\gamma : E_{\gamma(a)} \to E_{\gamma(b)}$ be the map $v \to s_v(b)$, where s_v is the parallel section of E along the curve γ for which $s_v(a) = v$. It is clear that T_γ is a linear map. It is called a **parallel transport along the curve** γ (from the point $\gamma(a)$ to the point $\gamma(b)$).

If σ is a parallel section of E along a curve γ, then, for every $a \leq t_1 < t_2 \leq b$, the result of the parallel transport of σ_{t_1} along the curve $\gamma|[t_1, t_2]$ is obviously σ_{t_2}, i.e., σ is invariant with respect to the parallel transport along γ. If s is a section of the bundle E defined in a neighbourhood of $\gamma([a, b])$, then $\sigma = \gamma \circ s$ is a section of E along the curve γ. Suppose s is parallel: $Ds = 0$. Then $\widetilde{D}_{\frac{d}{dt}} \sigma = D_{\gamma_*(\frac{d}{dt})} s = 0$, thus the section $\sigma = \gamma \circ s$ of E along γ is parallel with respect to D. It follows that the section s is invariant under the parallel transport along every curve in its domain. This justifies the name "parallel" for a section s satisfying the identity $Ds = 0$.

Recall that a map $\gamma : [a, b] \to M$ is called a piecewise smooth curve in M if there is a partition $a = t_0 < t_1 < ... < t_{k-1} < t_k = b$ of the interval $[a, b]$ such that each arc $\gamma|[t_{i-1}, t_i]$ is a smooth curve, $i = 1, ..., k$. The notion of parallel transport along a smooth curve can be extended to the case of piecewise smooth curves in an obvious way. If $\gamma_1 : [a_1, b_1] \to M$ and $\gamma_2 : [a_2, b_2] \to M$ are piecewise smooth curves with

$\gamma_1(b_1) = \gamma_2(a_2)$, their composition $\gamma = \gamma_2.\gamma_1$ obtained by joining γ_1 and γ_2 end to end is also a piecewise smooth curve, and clearly $T_\gamma = T_{\gamma_2} \circ T_{\gamma_1}$. In particular, the notion of parallel transport on a piecewise smooth curve does not depend on its partition into smooth arcs. If $\gamma : [a, b] \to M$ is a piecewise smooth curve, its reversely oriented curve starting from the end point $\gamma(b)$ of γ and going in direction to its initial point $\gamma(a)$ is given by the formula $\gamma^{-1}(t) = \gamma(a + b - t)$, $t \in [a, b]$. The composition $\alpha = \gamma^{-1}.\gamma$ is the constant curve $\alpha(t) = \gamma(a)$, and the composition $\beta = \gamma.\gamma^{-1}$ is the constant curve $\beta(t) = \gamma(b)$. The parallel transport on a constant curve is the identity map. Hence the linear map T_γ is invertible and its inverse map is $T_{\gamma^{-1}}$, the parallel transport along the reversely oriented curve. Of course, this follows from the fact that if s is a parallel section along γ, then $t \to s(a+b-t)$ is a parallel section along γ^{-1}.

Problem 2.11.1 Prove that any two points of a connected smooth manifold can be connected by a piecewise smooth curve.

As is well-known, the stronger statement that any two points of a connected smooth manifold can be connected by a smooth curve is also true. We shall show this fact as a corollary of a result that we will prove later when considering the problem of existence of parallel frames.

Proposition 2.11.2 *Let $E \to M$ be a vector bundle with connection D, $\gamma : J \to M$ a smooth curve in M, and s a smooth section of E along the curve γ. Let $s_{(t)}$ be the unique parallel section of E along γ with respect to D for which $s_{(t)}(t) = s(t)$. Then if $\widetilde{D}$ is the connection on the bundle γ^*E induced by D, and $t_0 \in J$, then*

$$\widetilde{D}_{\frac{d}{dt}} s|_{t_0} = \lim_{t \to t_0} \frac{s_{(t)}(t_0) - s(t_0)}{t - t_0}.$$

Proof Let $e_1, ..., e_r$ be a basis of $E_{\gamma(t_0)}$, and let $s_1, ..., s_r$ be the parallel sections of E along γ satisfying the initial condition $s_j(t_0) = e_j$, $j = 1, ..., r$. These sections constitute a basis of $E_{\gamma(t)}$ for every $t \in J$ since the parallel transport is an isomorphism and $s_1, ..., s_r$ are invariant under the parallel transport along γ. Hence $s_1, ..., s_r$ is a frame of the bundle γ^*E. Let $s = \sum_{i=1}^r f_i s_i$. By the linearity of the parallel transport, $s_{(t)}(t_0) = \sum_{i=1}^r f_i(t) s_i(t_0)$. Then $s_{(t)}(t_0) - s(t_0) = \sum_{i=1}^r [f_i(t) - f_i(t_0)] s_i(t_0)$. Therefore

$$\lim_{t \to t_0} \frac{s_{(t)}(t_0) - s(t_0)}{t - t_0} = \sum_{i=1}^r \frac{df_i}{dt}(t_0) s_i(t_0).$$

On the other hand,

$$\widetilde{D}_{\frac{d}{dt}} s = \sum_{i=1}^r \frac{df_i}{dt} s_i + \sum_{i=1}^r f_i \widetilde{D}_{\frac{d}{dt}} s_i = \sum_{i=1}^r \frac{df_i}{dt} s_i.$$

This proves the proposition. □

Corollary 2.11.1 *Let $E \to M$ be a vector bundle with connection D, s a section of E in a neighbourhood of a point $p \in M$, and $X \in T_pM$. Let $\gamma : J \to M$ be a smooth curve such that $\gamma_*(\frac{d}{dt}(t_0)) = X$ for a point $t_0 \in J$. Denote by $s_{(t)}$ the unique parallel section of E along γ with respect to D for which $s_{(t)}(t) = s(\gamma(t))$. Then*

$$D_X s = \lim_{t \to t_0} \frac{s_{(t)}(t_0) - s(t_0)}{t - t_0}.$$

Proof This identity follows from the preceding proposition and the identity $D_X s = \widetilde{D}_{\frac{d}{dt}}(s \circ \gamma)$. □

Note that $s_{(t)}(\tau)$ is the result of the parallel transport of $s(\gamma(t))$ along the arc of γ connecting the points $\gamma(t)$ and $\gamma(\tau)$. Therefore the notion of a connection (covariant derivative) can be described by means of the concept of parallel transport along a curve.

Corollary 2.11.2 *Let $E \to M$ be a vector bundle with connection D, s a section of E in a neighbourhood of a point $p \in M$, and $X \in T_pM$. Let $\gamma : J \to M$ be a smooth curve such that $\gamma_*(\frac{d}{dt}(t_0)) = X$ for a point $t_0 \in J$. Denote by $s_{(t)}$ the unique parallel section of E along γ with respect to D for which $s_{(t)}(t) = s(\gamma(t))$. If $s \circ \gamma$ is a parallel section along γ with respect D, then $D_X s = 0$.*

Let $\mathcal{H}$ be the horizontal bundle of E with respect to a connection D. By Lemma 2.10.1, if s is a parallel section of E along a smooth curve $\gamma : [a, b] \to M$, then $s : [a, b] \to E$ is a smooth curve such that $s_*(\frac{d}{dt}(\tau)) \in \mathcal{H}_{s(\tau)}$ and $\pi \circ s = \gamma$, where $\pi : E \to M$ is the projection map. A curve s with this property is called a horizontal lift of γ. Conversely, if $s : [a, b] \to E$ is a horizontal lift of γ, then it follows from Lemma2.10.1 that s is a parallel section of E along γ.

Let p, q be two points of M, and let $\gamma : [a, b] \to M$ be a smooth curve connecting p and q: $\gamma(a) = p, \gamma(b) = q$. In general, the parallel transport from p to q depends on the curve connecting the points. We will show that if the connection D is flat, the parallel transport along (smoothly) homotopic curves gives the same result. More precisely, let $h : [a, b] \times [c, d] \to M$ be a smooth map such that $h(t, c) = \gamma(t)$, $h(a, u) = \gamma(a)$, $h(b, u) = \gamma(b)$ for every u. Thus, $\gamma_u(t) = h(t, u)$ is a family of smooth curves with the same ends p and q, smoothly depending on the parameter u, and passing through the curve γ. We claim that $T_\gamma = T_{\gamma_u} : E_p \to E_q$ for every u.

By the smoothness condition on γ and h, there exist open intervals $J \supset [a, b]$ and $I \supset [c, d]$ such that γ and h can be extended to smooth maps on J and, respectively, $J \times I$. Let $v \in E_p$, and for every $u \in I$, let s^u be the parallel section of E along the curve $\gamma_u : J \to M$ for which $s^u(a) = v$. Then, as we have noted above, $s(t, u) = s^u(t)$ is a smooth section of h^*E on $J \times I$. Set $\widetilde{D} = h^*D$, $\widetilde{D}^u = \gamma_u^* D$, $\gamma_t(u) = h(t, u)$, $\widetilde{D}^t = \gamma_t^* D$. Also, it is convenient to set $X = \frac{\partial}{\partial t}$, $Y = \frac{\partial}{\partial u}$ and

$s^t = s(t, u)$. Then, by formula (2.4),

$$\widetilde{D}_{(X_t, 0_u)} s = \widetilde{D}^u_{X_t} s^u + \widetilde{D}^t_{0_u} s^t = \widetilde{D}^u_{X_t} s^u = 0.$$

If R is the curvature of the connection D and $\widetilde{R}$ is the curvature of the connection $\widetilde{D}$, then $\widetilde{R}((X, 0), (0, Y)s = R(h_*(X, 0), h_*(0, Y))s = 0$. Hence, since $[(X, 0), (0, Y)] = 0$,

$$\widetilde{D}_{(X,0)} \widetilde{D}_{(0,Y)} s = \widetilde{D}_{(0,Y)} \widetilde{D}_{(X,0)} s = \widetilde{D}_{(0,Y)} 0 = 0.$$

Denote the section $\widetilde{D}_{(0,Y)} s : J \times I \to h^* E$ of E along h by σ. Then for every u,

$$\widetilde{D}^u_{X_t} \sigma^u = \widetilde{D}_{(X_t, 0_u)} \sigma = \widetilde{D}_{(X_t, 0_u)} \widetilde{D}_{(0,Y)} s = 0.$$

Therefore the section σ^u is parallel along the curve γ_u. Furthermore, since $s^a(u) = v \equiv const$,

$$\sigma^u(a) = \widetilde{D}_{(0_a, Y_u)} s = \widetilde{D}^a_{Y_u} s^a = 0.$$

It follows that $\sigma^u \equiv 0$. Then

$$\widetilde{D}^b_Y s^b = \widetilde{D}_{(0,Y)} s = 0.$$

Since γ_b; $I \to M$ is the constant curve $\gamma_b(u) \equiv q$, the bundle $\gamma_b^* E$ is trivial and $\widetilde{D}^b = \gamma_b^* D$ is the trivial connection on it. Thus, the identity $\widetilde{D}^b_Y s^b = 0$ reads as $ds^b = 0$, hence $s^b \equiv const$. It follows that, for every $u \in I$,

$$T_\gamma(v) = s^c(b) = s(b, c) = s(b, u) = s^u(b) = T_{\gamma_u}(v).$$

This proves the following statement.

Proposition 2.11.3 *Let $E \to M$ be a vector bundle with a connection D. If $\gamma_0 : [a, b] \to M$ and $\gamma_1 : [a, b] \to M$ are smooth curves with the same ends $p = \gamma_o(a) = \gamma_1(a)$, $q = \gamma_o(b) = \gamma_1(b)$, that are smoothly homotopic, then the parallel transports*

$$T_{\gamma_0} : E_p \to E_q, \quad T_{\gamma_1} : E_p \to E_q$$

with respect to D coincide.

Note that if two smooth curves (and more generally, two smooth maps) are homotopic by a continuous homotopy, then they are smoothly homotopic, see, e.g., J. M. Lee [15], Chap. 6, Proposition 6.29.

As we have already mentioned, if there is a parallel frame of E in a neighbourhood of every point of M, the connection D is flat. Now, we are going to show that the converse statement is also true. For this, we need the following lemma.

Lemma 2.11.1 *Let M be a smooth manifold and p_0, p' two points of M. Let (U, φ) be a chart of the manifold M at p' such that $\varphi(U)$ is a ball in $\mathbb{R}^n$ with centre the* *point $a' = \varphi(p')$. If a point $p \in U$ can be connected with p_0 by a smooth curve, then there is a neighbourhood U' of p', $U' \subset U$, such that*

(i) $U' \ni p$.
(ii) for every point $q \in U'$ and every tangent vector $v \in T_qM$ at q, there is a smooth curve connecting p_0 with q and passing through the point p, whose tangent vector at the point q is the vector v and whose arc from p to q lies in U.

Proof The idea of the proof is to connect p_0 and p by a smooth curve γ and to take a point $p_1 \in U$ on γ after the point p (in the direction from p_0 to p). Since U is (diffeomorphic to) a ball, it is clear that if $q \in U$ and $v \in T_qM$, we can connect p_1 with q by a smooth curve c lying in U, whose tangent vector at q is v. The continuous curve $\widetilde{\alpha}$ consisting of γ and c passes through the points p_0, p, p_1, q and is not smooth only at the point p_1. We smooth this curve near the point p_1 so that the resulting smooth curve contains an arc of γ passing through p and an arc of c passing through q. The technical implementation of this idea is as follows.

Let $\gamma : (-\varepsilon, 1+\varepsilon) \to M$ be a smooth curve connecting p_0 with p: $\gamma(0) = p_0$, $\gamma(1) = p$. Denote the radius of the ball $\varphi(U)$ by r, and let B' be the ball with center a' and radius r', $r' < r$, such that the points a' and $a = \varphi(p)$ lie B'. Set $U' = \varphi^{-1}(B')$. Let ε', $0 < \varepsilon' < 1$, be such that $\gamma(1-\varepsilon', 1+\varepsilon') \subset U'$. Then the curve $\alpha(t) = \varphi \circ \gamma(t)$, $t \in (1-\varepsilon', 1+\varepsilon')$, is smooth, lies in the ball B', and passes through the point a. Fix a number $t_1 \in (1, 1+\varepsilon')$ and set $a_1 = \alpha(t_1)$. Take a point $q \in U'$ and a tangent vector $v \in T_qM$. Let $c : (t_1 - \delta, t_1 + 1 + \delta) \to B'$, $0 < \delta < t_1 - 1$, be a smooth curve in the ball B' for which $c(t_1) = a_1$, $c(t_1+1) = b$, and the derivative $\dot{c}\,(t_1+1) = \varphi_*(v)$. Set $t_2 = t_1 + 1$ and

$$\widetilde{\alpha}(t) = \begin{cases} \alpha(t) \text{ for } & 1-\varepsilon' < t \le t_1, \\ c(t) \text{ for } & t_1 \le t < t_2 + \delta \end{cases}$$

Let t' be a point of the interval $(1-\varepsilon', 1)$ and t'' a point of the interval $(t_2, t_2+\delta)$. By the Weierstrass theorem for uniform approximation of continuous functions by polynomials, there exists a smooth map $F : \mathbb{R} \to \mathbb{R}^n$ such that

$$||F(t) - \widetilde{\alpha}(t)|| < r - r'$$

for every $t \in [t', t'']$. Let δ' be such that $[t_1 - \delta', t_1 + \delta'] \subset (t_1 - \delta, t_1 + \delta)$. Thus, on the real line, we have the points

$$1 - \varepsilon' < t' < 1 < t_1 - \delta < t_1 - \delta' < t_1 < t_1 + \delta' < t_1 + \delta < t_2 < t'' < t_2 + \delta.$$

Let χ be a smooth function on the real line such that $0 \le \chi \le 1$, $\chi \equiv 0$ out of the interval $(t_1 - \delta, t_1 + \delta)$ and $\chi \equiv 1$ on the interval $(t_1 - \delta', t_1 + \delta')$ (a "bump" function). Set

$$\beta(t) = \chi(t)F(t) + (1 - \chi(t))\widetilde{\alpha}(t), \quad t \in (1 - \varepsilon', t_2 + \delta).$$

We have $\beta = F$ in the interval $(t_1 - \delta', t_1 + \delta')$, hence β is a smooth curve on this interval. The curve $\widetilde{\alpha}$ is smooth on the intervals $(1 - \varepsilon', t_1)$ and $(t_1, t_2 + \delta)$, hence β is also smooth on these intervals. Thus, the curve β is smooth on the intervals $(1-\varepsilon', t_1), (t_1, t_2+\delta), (t_1-\delta', t_1+\delta')$ whose union is $(1-\varepsilon', t_2+\delta)$. Note also that $\beta(t) = \widetilde{\alpha}(t)$ for $t \in (1-\varepsilon', t_1-\delta) \cup (t_1+\delta, t_2+\delta)$. Next, for every $t \in (1-\varepsilon', t_2+\delta)$,

$$\begin{aligned} ||\beta(t) - a|| &\le ||\beta(t) - \widetilde{\alpha}(t)|| + ||\widetilde{\alpha}(t) - a|| = \chi(t)||F(t) - \widetilde{\alpha}(t)|| + ||\widetilde{\alpha}(t) - a|| \\ &\le ||F(t) - \widetilde{\alpha}(t)|| + ||\widetilde{\alpha}(t) - a||. \end{aligned}$$

Therefore if $t \in [t', t'']$,

$$||\beta(t) - a|| < (r - r') + r' = r.$$

Hence $\beta(t) \in \varphi(U)$ for $t \in [t', t'']$. Let t_0 be a point of the interval $(1, t_1 - \delta)$. Note that for $t \in (1, t_1 - \delta)$, we have $\beta(t) = \widetilde{\alpha}(t) = \alpha(t) = \varphi \circ \gamma(t)$. Now, set

$$\widetilde{\gamma}(t) = \begin{cases} \gamma(t) \text{ for } & -\varepsilon < t \le t_0, \\ \varphi^{-1} \circ \beta(t) \text{ for } & t_0 \le t < t'' \end{cases}$$

The curve $\widetilde{\gamma}$ is smooth since $\widetilde{\gamma}(t) = \gamma(t)$ in the neighbourhood $(1, t_1-\delta)$ of the point t_0. Furthermore, $\widetilde{\gamma}(0) = \gamma(0) = p_0$, $\widetilde{\gamma}(1) = \gamma(1) = p$ and $\widetilde{\gamma}(t_2) = \varphi^{-1} \circ \beta(t_2) = \varphi^{-1} \circ \widetilde{\alpha}(t_2) = \varphi^{-1}(b) = q$. Moreover, in the neighbourhood $(t_1 + \delta, t'')$ of the point t_2, $\widetilde{\gamma}(t) = \varphi^{-1} \circ \widetilde{\alpha}(t) = \varphi^{-1} \circ c(t)$. Therefore $\widetilde{\gamma}_*(\frac{d}{dt}(t_2)) = \varphi_*^{-1} \circ \dot{c}\ (t_2) = \varphi_*^{-1} \circ \varphi_*(v) = v$. □

Corollary 2.11.3 *Every two points of a connected smooth manifold M can be connected by a smooth curve.*

Proof Let p_0 be a point of M. Denote by E the set of points of M that can be jointed with p_0 by a smooth curve. Clearly, $p_0 \in E$ (together with a neighbourhood). By Lemma 2.11.1, the sets E and $M \setminus E$ are open. Hence $E = M$. □

Lemma 2.11.1 and Corollary 2.11.3 imply the following

Corollary 2.11.4 *Let M be a connected smooth manifold, p_0 and p two points of M, and $v \in T_pM$ a tangent vector. Then there is a smooth curve in M connecting the points p_0 and p whose tangent vector at the point p is the vector v.*

This and the proof of Lemma 2.11.1 imply

Corollary 2.11.5 *Under the notation of the preceding corollary, if $v_0 \in T_{p_0}M$ is a tangent vector, then there is a smooth curve connecting p_0 and p whose tangent vectors at p_0 and p are v_0 and v, respectively.*

Proposition 2.11.4 *Let $E \to M$ be a vector bundle with a connection D. For a point $p \in M$, let $a_1, ..., a_r$ be a basis of the fibre E_p. Suppose that the connection D is flat and the manifold M is simply connected. Then there is a globally defined frame of sections $e_1, ..., e_r$ of E, parallel with respect to D and such that $e_i(p) = a_i$, $i = 1, .., r$.*

Proof Fix a point $p_0 \in M$. For every point $p \in M$, take a smooth curve γ connecting p_0 with p, and set $e_i = T_\gamma(a_i)$, the parallel transport of a_i along the curve γ to an element of the fibre E_p, $i = 1, ..., r$. Any curve γ' connecting p_0 with p is continuously homotopic to γ. The curves γ and γ' are then smoothly homotopic, therefore, by Proposition 2.11.3, $T_{\gamma'}(a_i) = T_\gamma(a_i)$, i.e., $e_i(p)$ does not depend on the choice of the curve γ. The parallel transport is an isomorphism of fibres, hence $e_1(p), ..., e_r(p)$ is a basis of E_p for every p. To show that the sections $e_1, ..., e_r$ of E are smooth, take any point p' of M. Let (U, φ) be a chart of M at p' for which $\varphi(U)$ is a ball with center $\varphi(p')$. By Lemma 2.11.1, there is a neighbourhood $U' \subset U$ of p' such that the point p_0 can be connected with every point $q \in U'$ by a smooth curve γ_q passing through the point p' and such that the arc of γ_q from p' to q lies in U. Let γ_q' and γ_q'' be the arcs of γ_q connecting p_0 with p' and p' with q, respectively. Then $e_i(q) = T_{\gamma_q}(a_i) = T_{\gamma_q''} \circ T_{\gamma_q'}(a_i) = T_{\gamma_q''}(e_i(p'))$. For every $q \in U'$, $\mu_q(t) = \varphi^{-1}((1-t)\varphi(p') + t\varphi(q))$, $0 \leq t \leq 1$, is a smooth curve in U connecting p' with q. We can reparametrize the curve γ_q'' by an affine change of the parameter so that its new parameter is varied in the interval $[0, 1]$. Since the parallel transport along a curve is invariant under affine changes of the curve parameter, we may assume that the defining interval of γ_q'' is $[0, 1]$. Since $\varphi(U)$ is a ball, the curves γ_q'' and μ_q are smoothly homotopic. Therefore $e_i(q) = T_{\gamma_q''}(e_i(p')) = T_{\mu_q}(e_i(p'))$. Let s_i^q be the parallel section of E along the curve μ_q for which $s_i^q(0) = e_i(p')$. Since the curve μ_q smoothly depends on the "parameter" q, the map $[0, 1] \times U \to E$, $(t, q) \to s_i^q(t)$, is smooth. Hence $q \to s_i^q(1)$ is a smooth map of U into E, i.e., e_i is a smooth section of E. Let X be a tangent vector of M at a point p. By Corollary 2.11.4, there is a smooth curve $\gamma : (-\varepsilon, 1 + \varepsilon) \to M$ such that $\gamma(0) = p_0$, $\gamma(1) = p$, $\gamma_*(\frac{d}{dt}(1)) = X$. By the very definition of e_i, the section $e_i \circ \gamma$ is parallel along the curve γ, hence $D_X e_i = 0$ by Corollary 2.11.2. This shows that the section e_i is parallel, and the proposition is proved. □

Corollary 2.11.6 *A vector bundle over a simply connected manifold admitting a flat connection is isomorphic to the trivial bundle endowed with the trivial connection.*

Problem 2.11.2 Show that if $e_1, ..., e_r$ and $e_1', ..., e_r'$ are parallel frames on open sets U and U' whose intersection is connected, the transition functions from one of the frames to the other are constants.

3 Tautological (Universal) Bundles

3.1 Complex Projective Spaces

Let V be a complex vector space of dimension $n+1 \geq 2$. The group $\mathbb{C}_* = \mathbb{C}\setminus\{0\}$ acts on the open set $V_* = V \setminus \{0\}$ by multiplication: $(\lambda, z) \to \lambda z$, $\lambda \in \mathbb{C}_*$, $z \in V_*$. Recall that the quotient topological space $\mathbb{P}(V) = V_*/\mathbb{C}_*$ is called the **projective space** of V. Every equivalence class $[z]$ is uniquely determined by a non-zero vector z of V. Every complex line of V (a 1-dimensional complex subspace) is also determined by such a vector. Hence the map $[z] \to \mathbb{C}.z$ is a bijection of $\mathbb{P}(V)$ onto the set of complex lines of V.

The projective space of a real vector space is defined in a similar way.

In the case $V = \mathbb{C}^{n+1}$, the projective space $\mathbb{P}(V)$ is usually denoted by $\mathbb{CP}^n$ or just by $\mathbb{P}^n$ (if it is clear that we are talking about the complex projective space). Let $S^{2n+1} = \{z = (z_0, \dots, z_n) \in \mathbb{C}^{n+1} : \sum_{k=1}^{n} |z_k|^2 = 1\}$ be the unit sphere. The group S^1 acts on S^{2n+1} by coordinate-wise multiplication: $(\lambda, z) \to \lambda.z = (\lambda z_0, \dots, \lambda z_n)$, $\lambda \in S^1$, $z \in S^{2n+1}$. The quotient space S^{2n+1}/S^1 is homeomorphic to $\mathbb{CP}^n$ via the map $S^{2n+1}/S^1 \ni [z] \to [z] \in \mathbb{C}_*^{n+1}/\mathbb{C}_* = \mathbb{CP}^n$; its inverse map is given by $[z] \to \left[\frac{z}{|z|}\right]$.

We show now that the quotient space S^{2n+1}/S^1 is Hausdorff. To this end, we will first establish that the natural projection $p : S^{2n+1} \to S^{2n+1}/S^1$ is an open map (sends open sets to open sets). Let U be an open subset of S^{2n+1}. The set $p(U)$ is open in the quotient topology exactly when the set $p^{-1}(p(U))$ is open. Let $w_0 \in p^{-1}(p(U))$. Then $w_0 = \lambda_0 z_0$ for some $z_0 \in U$ and $\lambda_0 \in S^1$. The map $\Phi : S^{2n+1} \to S^{2n+1}$ defined by $\Phi(w) = \lambda_0^{-1} w$ is continuous and $\Phi(w_0) = z_0 \in U$. Therefore $\Phi^{-1}(U)$ is an open neighbourhood of w_0. For every $w \in \Phi^{-1}(U)$, the point $z = \lambda_0^{-1} w \in U$, hence $w \in p^{-1}(p(U))$. Thus, $\Phi^{-1}(U) \subset p^{-1}(p(U))$. This shows that the set $p^{-1}(p(U))$ is open. Let R be the equivalence relation defined by the action of S^1 onto S^{2n+1}, i.e., $R = \{(z, w) \in S^{2n+1} \times S^{2n+1} : w = \lambda z \text{ for some } \lambda \in S^1\}$.

J. Davidov, *Vector Bundles and Connections*, Compact Textbooks in Mathematics,
https://doi.org/10.1007/978-3-032-07403-4_3

Taking into account the compactness of S^1, it is easy to see that if $R \ni (z_\nu, w_\nu) \to (z, w) \in S^{2n+1} \times S^{2n+1}$ when $\nu \to \infty$, then $(z, w) \in R$. Therefore the set R is closed. Then if $p(z) \neq p(w)$, the point (z, w) lies in the open set $(S^{2n+1} \times S^{2n+1}) \setminus R$. Hence the points z and w possess neighbourhoods U and V in S^{2n+1} such that $U \times V \subset (S^{2n+1} \times S^{2n+1}) \setminus R$. Since the natural projection is an open map, the sets $p(U)$ and $p(V)$ are disjoint neighbourhoods of $p(z)$ and $p(w)$. Therefore S^{2n+1}/S^1 is a Hausdorff topological space.

Another proof of this fact will be given later when considering the Grassmann manifolds.

Note that the space S^{2n+1}/S^1 is compact and connected as the image of the compact and connected space S^{2n+1} under the continuous map p

Therefore the space $\mathbb{CP}^n$ is Hausdorff, compact, and connected.

The space $\mathbb{CP}^n$ can be covered by the open sets

$$U_j = \{[z_0, \ldots, z_n] \in \mathbb{CP}^n : z_j \neq 0\}, \quad j = 0, 1, \ldots, n. \tag{3.1}$$

Each of these sets is homeomorphic to $\mathbb{C}^n$ via the map

$$\varphi_j([z_o, \ldots, z_n]) = (\frac{z_0}{z_j}, \ldots, \widehat{\frac{z_j}{z_j}}, \ldots, \frac{z_n}{z_j}); \tag{3.2}$$

the inverse map is clearly given by

$$\varphi_j^{-1}(\zeta_0, \ldots, \zeta_{n-1}) = [\zeta_0, \ldots, \zeta_{j-1}, 1, \zeta_j, \ldots, \zeta_{n-1}].$$

Here, as usual, the symbol $\widehat{\ }$ means that the object over which it is placed is omitted. If $k < j$, we have $\varphi_j(U_j \cap U_k) = \{(\zeta_0, \ldots, \zeta_{n-1}) \in \mathbb{C}^n : \zeta_k \neq 0\}$ and

$$\varphi_k \circ \varphi_j^{-1}(\zeta_0, \ldots, \zeta_{n-1}) = (\frac{\zeta_0}{\zeta_k}, \ldots, \widehat{\frac{\zeta_k}{\zeta_k}}, \ldots, \frac{\zeta_{j-1}}{\zeta_k}, \frac{1}{\zeta_k}, \frac{\zeta_j}{\zeta_k}, \ldots, \frac{\zeta_{n-1}}{\zeta_k}).$$

If $k > j$, then $\varphi_j(U_j \cap U_k) = \{(\zeta_0, \ldots, \zeta_{n-1}) \in \mathbb{C}^n : \zeta_{k-1} \neq 0\}$ and

$$\varphi_k \circ \varphi_j^{-1}(\zeta_0, \ldots, \zeta_{n-1}) = (\frac{\zeta_0}{\zeta_{k-1}}, \ldots, \frac{\zeta_{j-1}}{\zeta_{k-1}}, \frac{1}{\zeta_{k-1}}, \frac{\zeta_j}{\zeta_{k-1}}, \ldots, \widehat{\frac{\zeta_{k-1}}{\zeta_{k-1}}}, \ldots, \frac{\zeta_{n-1}}{\zeta_{k-1}}).$$

Therefore the maps $\varphi_k \circ \varphi_j^{-1}$ are holomorphic and $\{U_j, \varphi_j\}$ is a complex-analytic atlas on $\mathbb{CP}^n$. It defines the structure of a complex manifold of dimension n on the projective space.

The open sets U_j possess a countable base of their topology (as homeomorphic to $\mathbb{C}^n$) and constitute a finite cover of the projective space. Hence this space has a countable base of the topology. Thus, $\mathbb{CP}^n$ is a Hausdorff, connected manifold with a countable base, hence paracompact.

Let $p : \mathbb{C}_*^{n+1} \to \mathbb{C}_*^{n+1}/\mathbb{C}_* = \mathbb{CP}^n$ be the natural projection. The map p sends the open set $V_j = \{(\zeta_0, \ldots, \zeta_n) \in \mathbb{C}_*^{n+1} : \zeta_j \neq 0\}$ onto the set U_j, the domain of the map φ_j, and $\varphi_j \circ p\,(\zeta_0, \ldots, \zeta_n) = (\frac{\zeta_0}{\zeta_j}, \ldots, \widehat{\frac{\zeta_j}{\zeta_j}}, \ldots, \frac{\zeta_n}{\zeta_j})$. The Jacobi matrix of this holomorphic map is of rank n. Hence $rank\ p_* = n$, i.e., the natural projection p is a submersion.

Every basis $e = (e_0, \ldots, e_n)$ of V determines a linear isomorphism $\mathbb{C}^{n+1} \cong V$ which yields a homeomorphism $h_e : \mathbb{CP}^n \to \mathbb{P}(V)$ defined by $h_e([z_0, \ldots, z_n]) = [z_0 e_0 + \cdots + z_n e_n]$. Using the homeomorphisms h_e, the complex structure of $\mathbb{CP}^n$ can be transferred to $\mathbb{P}(V)$. Thus, on $\mathbb{P}(V)$ we have an atlas $\{U_j, \varphi_j\}$ defined by formulas (3.1) and (3.2), in which $z_0, \ldots, z_n$ are the coordinates of a point z of V_* with respect to the basis $e_0, \ldots, e_n$. Let $e' = (e'_0, \ldots, e'_n)$ be another basis of V and $\{U'_j, \varphi'_j\}$ its corresponding atlas on $\mathbb{P}(V)$. Set $e_p = \sum_{q=0}^n a_{pq} e'_q$, $p = 0, 1, \ldots, n$, $a_{pq} \in \mathbb{C}$. Then

$$\varphi_j(U_j \cap U'_k) = \{(\zeta_0, \ldots, \zeta_{n-1}) \in \mathbb{C}^n : \sum_{l=0}^{j-1} \zeta_l a_{lk} + a_{jk} + \sum_{l=j}^{n-1} \zeta_l a_{l+1,k} \neq 0\}.$$

In particular, the set $\varphi_j(U_j \cap U'_k)$ is open. It is easy to see that

$$\varphi'_k \circ \varphi_j^{-1}(\zeta_0, \ldots, \zeta_{n-1}) = (\frac{z'_0}{z'_k}, \ldots, \widehat{\frac{z'_k}{z'_k}}, \ldots, \frac{z'_n}{z'_k}),$$

where

$$z'_m = \sum_{l=0}^{j-1} \zeta_l a_{lm} + a_{jm} + \sum_{l=j}^{n-1} \zeta_l a_{l+1,m}, \quad m = 0, \ldots, n.$$

This shows that the maps $\varphi'_k \circ \varphi_j^{-1}$ are holomorphic. Therefore the atlases on $\mathbb{P}(V)$ defined by means of the bases e and e' determine the same complex structure.

Similar considerations can be applied for real projective spaces. They have the structure of real-analytic manifolds.

Consider the sphere S^2 with its standard structure of a complex manifold. Recall that this structure is defined in the following way. Let $N = (0, 0, 1)$ and $S = (0, 0, -1)$ be the north and south poles of the sphere. Set $V_0 = S^2 \setminus \{N\}$ and $V_1 = S^2 \setminus \{S\}$. Let $\psi_0 : V_0 \to \mathbb{R}^2 = \mathbb{C}$ and $\psi_1 : V_1 \to \mathbb{R}^2 = \mathbb{C}$ be the stereographic projections. Then $\psi_0(V_0 \cap V_1) = \mathbb{C} \setminus \{0\}$ and

$$\psi_1 \circ \psi_0^{-1}(\zeta) = \frac{1}{\zeta}, \quad \zeta \in \mathbb{C} \setminus \{0\}. \tag{3.3}$$

Hence $\{(V_0, \psi_0), (V_1, \overline{\psi}_1)\}$ is a complex-analytic atlas that defines the structure of a complex manifold on S^2.

Problem 3.1.1 Let J be the complex structure of S^2. Identify the tangent space $T_a S^2$ with the space of vectors v in $\mathbb{R}^3$ orthogonal to a. Show that $Jv = -a \times v$, where $\times$ is the standard cross product of vectors in $\mathbb{R}^3$.

Proposition 3.1.1 *(i)* $\mathbb{RP}^1$ *is diffeomorphic to* S^1.
(ii) $\mathbb{CP}^1$ *is biholomorphic to* S^2.

Proof We give a proof only of the second statement, the first one can be proved in a similar way.

Let $\{(U_0, \varphi_0), (U_1, \varphi_1)\}$ be the standard atlas of $\mathbb{CP}^1$. Then $\varphi_0(U_0 \cap U_1) = \mathbb{C} \setminus \{0\}$ and

$$\varphi_1 \circ \varphi_0^{-1}(\zeta) = \frac{1}{\zeta}, \quad \zeta \in \mathbb{C} \setminus \{0\}.$$

This and identity (3.3) imply $\overline{\psi}_1 \circ \psi_0^{-1} = \varphi_1 \circ \varphi_0^{-1}$. It follows that the maps

$$\psi_0^{-1} \circ \varphi_0 : U_0 \to V_0 \text{ and } \overline{\psi}_1^{-1} \circ \varphi_1 : U_1 \to V_1$$

coincide on $U_0 \cap U_1$. Therefore these maps determine a globally defined holomorphic and bijective map $\mathbb{CP}^1 \to S^2$. The inverse map is clearly also holomorphic. □

Problem 3.1.2 Let V be a complex vector space and $g : V \to V$ a linear map. Prove that the map $\widetilde{g} : \mathbb{P}(V) \to \mathbb{P}(V)$ defined by $\widetilde{g}([v]) = [g(v)]$ is holomorphic.

3.2 The Tautological Bundle $\mathcal{O}(-1)$ and the Bundles $\mathcal{O}(k)$ over a Complex Projective Space

Let V be a complex vector space, $dim\ V = n + 1 \geq 2$, and let $\mathbb{P}(V)$ be its projective space. There is a natural holomorphic line vector bundle over $\mathbb{P}(V)$; its fibre over a point $[z]$ is just the line $\mathbb{C}.z$. This bundle is called **tautological** and is denoted by $\mathcal{O}(-1)$ (the presence of -1 in the notation will become clear in Chap. 4 where the Chern classes are discussed). Thus,

$$\mathcal{O}(-1) = \{([z], \xi) \in \mathbb{P}(V) \times V : \xi \in \mathbb{C}.z\}.$$

The tautological bundle is also called universal. The reason for this name will be explained in Sect. 3.4.

Let h be a Hermitian metric on V. Every equivalence class of $\mathbb{P}(V)$ has a representative with $|z|_h = 1$. With this trivial remark in mind, it is easy to see that $\mathcal{O}(-1)$ is a closed subset of $\mathbb{P}(V) \times V$. We endow $\mathcal{O}(-1)$ with the induced topology. Let $\pi : \mathcal{O}(-1) \to \mathbb{P}(V)$ be the map $\pi([z], \xi]) = [z]$. Then $\pi^{-1}([z]) = ([z], \mathbb{C}.z)$ has an obvious structure of a 1-dimensional complex vector space that does not depend on the choice of z. Take a basis $e_0, \ldots, e_n$ of V and, as above, set $U_j = \{[z_0e_0 + \cdots + z_ne_n] \in \mathbb{P}(V) : z_j \neq 0\}$, $j = 0, \ldots, n$. For every $k = 0, \ldots, n$ with $k \neq j$, set $s_k([z]) = ([z], e_k)$. Furthermore, let $s_j([z]) = ([z], \frac{1}{z_j}z)$. Then $s_0, \ldots, s_n$ is a holomorphic frame of $\mathbb{P}(V) \times V$ on U_j, and the values of the section s_j lie in $\mathcal{O}(-1)$. Therefore, by Chap. 1, Proposition 1.7.3, $\mathcal{O}(-1)$ possesses a natural structure of a holomorphic subbundle of the trivial bundle $\mathbb{P}(V) \times V$ for which each section s_j is a frame. Let $g_{ij} : U_i \cap U_j \to GL(\mathbb{C}) \cong \mathbb{C}_*$ be the transition functions of the bundle $\mathcal{O}(-1)$ with respect to the trivializations determined by the sections s_j. Then $s_j = s_i.g_{ij}$ (see Chap. 1, Sect. 1.6). Obviously, $s_j = \dfrac{z_i}{z_j} s_i$. Hence $g_{ij}([z]) = \dfrac{z_i}{z_j}$.

Hereafter, we denote the trivial bundle $\mathbb{P}(V) \times V$ by $\underline{V}$. We denote the quotient bundle $\underline{V}/\mathcal{O}(-1)$ by Q; it is a holomorphic vector bundle over $\mathbb{P}(V)$ of rank n. Thus, the sequence

$$0 \longrightarrow \mathcal{O}(-1) \overset{inclusion}{\longrightarrow} \underline{V} \overset{projection}{\longrightarrow} Q \longrightarrow 0$$

is exact. The dual bundle of the tautological bundle $\mathcal{O}(-1)$ is denoted by $\mathcal{O}(1)$. It is called a **hyperplane** bundle; the reason for this name will become clear in Chap. 4.

The exactness of the sequence above implies that the sequence

$$0 \longrightarrow Q^* \longrightarrow \underline{V}^* \longrightarrow \mathcal{O}(1) \longrightarrow 0$$

is also exact.

For every natural number m, set

$$\mathcal{O}(-m) = \otimes^m \mathcal{O}(-1), \quad \mathcal{O}(m) = \otimes^m \mathcal{O}(1).$$

Set also $\mathcal{O}(0) = \underline{V}$.

The following statement gives a description of the vector space

$$H^0(\mathbb{P}(V), \mathcal{O}(k)), \quad k \text{ an integer},$$

of global holomorphic sections of the bundle $\mathcal{O}(k)$.

Proposition 3.2.1 (1) *For any $k \geq 0$, there is a canonical isomorphism*

$$H^0(\mathbb{P}(V), \mathcal{O}(k)) \cong \odot^k V^*.$$

(2) *If $k < 0$, then $H^0(\mathbb{P}(V), \mathcal{O}(k)) = 0$.*

Proof (1) Let $k \geq 0$. Every element of $\odot^k V^*$ can be considered as a multilinear symmetric form $a : \underbrace{V \times \cdots \times V}_{k} \to \mathbb{C}$. Let $[z] \in \mathbb{P}(V)$. The fibre of the bundle $\mathcal{O}(-1)$ over $[z]$ is the line $\mathbb{C}.z$ in V, and we denote the restriction of a to $\underbrace{\mathbb{C}.z \times \cdots \times \mathbb{C}.z}_{k}$ by $\widetilde{a}_{[z]}$. The resulting form can be viewed as an element of the space $\otimes^k (\mathbb{C}.z)^* = \otimes^k \mathcal{O}(-1)^*_{[z]}$, i.e., as an element of the fibre of the bundle $\mathcal{O}(k)$ over $[z]$. The map $\sigma_a : \mathbb{P}(V) \to \mathcal{O}(k)$ defined by $[z] \to \widetilde{a}_{[z]}$ is a holomorphic section of $\mathcal{O}(k)$. Indeed, let $[v] \in \mathbb{P}(V)$ and take a basis $e_0, \ldots, e_n$ of V with $e_0 = v$. We denote the coordinates of the points z of V with respect to this basis by $z_0, \ldots, z_n$.

The point [v] lies in the set $U_0 = \{[z_0, \dots, z_n] : z_0 \neq 0\}$, and $s([z]) = ([z], \frac{1}{z_0} z)$ is a frame of $\mathcal{O}(-1)$ over U_0. If s^* is the dual frame, $\otimes^k s^*$ is a frame of $\mathcal{O}(k)$ and $\sigma_a = h(\otimes^k s^*)$ on U_0, where $h([z]) = \sigma_a([z])(s([z]), \dots, s([z])) = a(\frac{z}{z_0}, \dots, \frac{z}{z_0})$. The latter identity shows that the function h is holomorphic. It follows that the section σ_a is holomorphic.

Conversely, let σ be a global holomorphic section of $\mathcal{O}(k)$, $k \geq 0$. For every $z \in V_*$, the value $\sigma([z])$ of σ can be considered as a multilinear form $\mathcal{O}(-1)_{[z]} \times \dots \times \mathcal{O}(-1)_{[z]} \to \mathbb{C}$. Since $\beta(z) = ([z], z)$ is a basis of $\mathcal{O}(-1)_{[z]}$, the form $\sigma([z])$ is uniquely determined by the number

$$f(z) = \sigma([z])(\underbrace{\beta(z), \dots, \beta(z)}_{k}).$$

In this way, we obtain a function $f : V_* \to \mathbb{C}$. This function is holomorphic. Indeed, let $v \in V_*$ and, as above, $e_0 = v, e_1, \dots, e_n$ be a basis of V. Then v lies in the open set $V_0 = \{\sum_{k=0}^n z_k e_k : z_0 \neq 0\}$. Clearly, if $z \in V_0$, the point $[z] \in U_0 = \{[z_0, \dots, z_n] : z_0 \neq 0\}$. Let s be the frame of $\mathcal{O}(-1)$ over U_0 defined by $s([z]) = ([z], \frac{1}{z_0} z)$. We have $s([z]) = \frac{1}{z_0}\beta(z)$, therefore $\sigma([z]) = \frac{1}{z_0^k} f(z) \otimes^k s^*([z])$ for $z \in V_0$. It follows that the function $\frac{1}{z_0^k} f(z)$ is holomorphic on V_0, hence the function f is holomorphic V_0. Thus, f is a holomorphic function on $V \setminus \{0\}$. Since $dim\ V \geq 2$, the function f extends to a holomorphic function on the whole space V (Hartogs' extension theorem, see, e.g., Griffiths and Harris [9], Chap. 0, Sect. 1). We denote the extension again by f. Since $\beta(\lambda z) = \lambda\beta(z)$ for $\lambda \in \mathbb{C}_*$, $f(\lambda z) = \lambda^k f(z)$ for $z \in V \setminus \{0\}$. Due to the continuity of f, this identity holds on the whole space V. Taking a basis of V, we identify V with $\mathbb{C}^{n+1}$. For $z = (z_0, \dots, z_n) \in \mathbb{C}^{n+1}$ and $\alpha = (\alpha_0, \dots, \alpha_n)$ where $\alpha_j \geq 0$ are integers, set $z^\alpha = z_0^{\alpha_0} \dots z_n^{\alpha_n}$. Since the function f is holomorphic on $\mathbb{C}^{n+1}$, it can be uniquely represented as the sum of a convergent power series

$$f(z) = \sum_\alpha c_\alpha z^\alpha, \quad c_\alpha \in \mathbb{C}.$$

The identity $f(\lambda z) = \lambda^k f(z)$, $\lambda \in \mathbb{C}_*$, is equivalent to the identity $\lambda^{\alpha_0 + \dots + \alpha_n} c_\alpha = \lambda^k c_\alpha$. Set $|\alpha| = \alpha_0 + \dots + \alpha_n$ and write the latter identity as $(\lambda^{|\alpha|-k} - 1)c_\alpha = 0$. Then $c_\alpha = 0$ for every α with $|\alpha| \neq k$. Thus, f is of the form $f(z) = \sum_{|\alpha|=k} c_\alpha z^\alpha$, i.e., f is a homogeneous polynomial of degree k. As is well-known, the space of homogeneous polynomials of degree k of $n+1$ complex variables is isomorphic to the space of multilinear symmetric forms $\underbrace{\mathbb{C}^{n+1} \times \dots \times \mathbb{C}^{n+1}}_{k} \to \mathbb{C}$ (see, e.g., Chap. 4, Sect. 2).

It follows that there is a unique element $a \in \odot^k V^*$ such that $f(z) = a(z, \dots, z)$ for every$z \in V$. For $z \in V \setminus \{0\}$, we have $\widetilde{a}_{[z]}(\beta(z), \dots, \beta(z)) = a(z, \dots, z) =$

$f(z) = \sigma([z])(\beta(z), \ldots, \beta(z))$. Since $\beta(z)$ is a basis of the one-dimensional space $\mathcal{O}(-1)_{[z]}$, we conclude that $\widetilde{a}_{[z]} = \sigma([z])$. Thus, $\sigma = \sigma_a$.
(2) Let σ be a global holomorphic section of $\mathcal{O}(-k)$, $k \geq 1$. For every $z \in V \setminus \{0\}$, $\otimes^k \beta(z)$ is a basis of the one-dimensional vector space $\otimes^k \mathcal{O}(-1)_{[z]} = \mathcal{O}(-k)_{[z]}$. Hence there is a unique complex number $f(z)$ such that $\sigma([z]) = f(z) \otimes^k \beta(z)$. Similarly to the considerations above, one can show that the function f is holomorphic on $V \setminus \{0\}$, hence it extends to a holomorphic function on V. If $\lambda \in \mathbb{C} \setminus \{0\}$ and $z \in V \setminus \{0\}$, then $f(\lambda z)\lambda^k \otimes^k \beta(z) = \sigma([\lambda z]) = \sigma([z]) = f(z) \otimes^k \beta(z)$, hence $f(\lambda z) = \lambda^{-k} f(z)$. Since f is continuous at the point 0, the latter identity holds for every $z \in V$. Then, if $f(z) \neq 0$ for some $z \in V$, we would have $f(0) = \lim_{\lambda \to 0} f(\lambda z) = \lim_{\lambda \to 0} (\lambda^{-k} f(z)) = \infty$, a contradiction. Therefore $f \equiv 0$, thus $\sigma = 0$. ◻

Proposition 3.2.2 *There is a canonical isomorphism*

$$T^h\mathbb{P}(V) \cong Q \otimes \mathcal{O}(1) \cong Hom(\mathcal{O}(-1), Q).$$

Proof Let $p : V_* = V \setminus \{0\} \to \mathbb{P}(V) = V_*/\mathbb{C}_*$ be the natural projection. Recall that for every $x \in V_*$ the tangent space $T_x^h V_*$ is canonically isomorphic to the vector space V via the map assigning $v \in V$ to the tangent vector in $T_x^h V_*$ that acts on (local) holomorphic functions ψ at x as the derivative of ψ at x in direction v: $v(\psi) = \frac{d}{d\zeta}\psi(x + \zeta v)|_{\zeta=0}$. Under this isomorphism, if χ is a holomorphic function in a neighbourhood of a point [x] of $\mathbb{P}(V)$,

$$p_*(v)(\chi) = v(\chi \circ p) = \frac{d}{d\zeta}(\chi \circ p(x + \zeta v))|_{\zeta=0}, \tag{3.4}$$

where ζ is a complex variable in a neighbourhood of the origin. Since $p(x + \zeta x) = p((1 + \zeta)x) = p(x)$ for every ζ sufficiently closed to 0, clearly $p_{*x}(x) = 0$. Hence $\mathbb{C}.x \subset Ker\ p_{*x}$. As we have noticed in Sect. 3.1, $dim\ Im\ p_{*x} = rank\ p_{*x} = n$, therefore $dim\ Ker\ p_{*x} = dim\ V - dim\ Im\ p_{*x} = 1$. Thus, $Ker\ p_{*x} = \mathbb{C}.x$, and p_{*x} induces an injective linear map $\widetilde{p}_{*x} : V/\mathbb{C}.x \to T^h_{[x]}\mathbb{P}(V)$. The space $V/\mathbb{C}.x$ is the fibre at the point $[x]$ of the bundle $\underline{V}/\mathcal{O}(-1) = Q$. Moreover $dim\,(V/\mathbb{C}.x) = n = dim\ T^h_{[x]}\mathbb{P}(V)$. Hence $\widetilde{p}_{*x} : Q_{[x]} \to T^h_{[x]}\mathbb{P}(V)$ is an isomorphism. This isomorphism, however, depends on the point x of $V \setminus \{0\}$ rather than the point $[x]$ of $\mathbb{P}(V)$. To see how p_{*x} depends on the choice of x on the line $[x] = \mathbb{C}.x$, we apply formula (3.4). If $\lambda \in \mathbb{C} \setminus \{0\}$, then $p(\lambda x + \zeta v) = p(x + \lambda^{-1}\zeta v)$ and (3.4) implies $p_{*\lambda x}(v) = \lambda^{-1} p_{*x}(v)$. Hence if $q \in Q_{[x]}$, we have $\widetilde{p}_{*\lambda x}(q) = \lambda^{-1}\widetilde{p}_{*x}(q)$. Set $\beta(x) = ([x], x)$. Then $\widetilde{p}_{*\lambda x}(q) \otimes \beta(\lambda x) = \widetilde{p}_{*x}(q) \otimes \beta(x)$. Therefore the map $\varphi_{[x]}(q) = \widetilde{p}_{*x}(q) \otimes \beta(x) : Q_{[x]} \to T^h_{[x]}\mathbb{P}(V) \otimes \mathcal{O}(-1)_{[x]}$ depends only on $[x]$. In this way, we obtain a map $\varphi : Q \to T^h\mathbb{P}(V) \otimes \mathcal{O}(-1)$, and we will show that this map is a (holomorphic) isomorphism of bundles. To this end, fix a point $[v] \in \mathbb{P}(V)$, and take a basis $e_0, e_1, \ldots, e_n$ of V. One of the coordinates $v_0, v_1, \ldots, v_n$ of v with respect to $e_0, e_1, \ldots, e_n$ is not zero, and, reordering the basis if necessary, we can assume that $v_0 \neq 0$.. Then $[v] \in U_0 = \{[z_0e_0 + \cdots + z_ne_n] \in \mathbb{P}(V) : z_0 \neq 0\}$.

For $x = \sum_{k=0}^{n} z_k e_k \in V$ with $z_0 \neq 0$, set $\varepsilon_0([x]) = ([x], \frac{x}{z_0})$, $\varepsilon_l([x]) = ([x], e_l)$, $l = 1, 2, \ldots, n$. Then $\varepsilon_0, \varepsilon_1, \ldots, \varepsilon_n$ is a holomorphic frame of the trivial bundle $\underline{V} = \mathbb{P}(V) \times V$ over U_0 and ε_0 is a frame of $\mathcal{O}(-1)$. Therefore if $pr : \underline{V} \to \underline{V}/\mathcal{O}(-1) = Q$ is the natural projection, $q_1 = pr \circ \varepsilon_1, \ldots, q_n = pr \circ \varepsilon_n$ is a holomorphic frame of Q over U_0. By the very definition of $\widetilde{p}_{*x}$, we have the following commutative diagramme for $[x] \in U_0$:

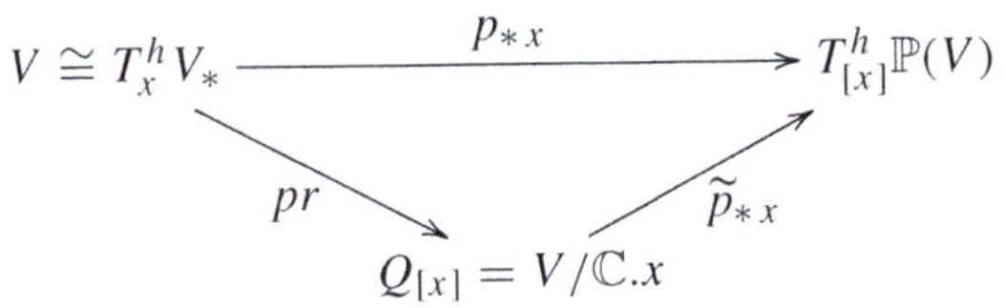

$$\begin{array}{ccc} V \cong T_x^h V_* & \xrightarrow{\quad p_{*x} \quad} & T_{[x]}^h \mathbb{P}(V) \\ & {\scriptstyle pr} \searrow \qquad \nearrow {\scriptstyle \widetilde{p}_{*x}} & \\ & Q_{[x]} = V/\mathbb{C}.x & \end{array}$$

Let $\zeta_l = \frac{z_l}{z_0}$, $l = 1, \ldots, n$, be the standard local coordinates of $\mathbb{P}(V)$ on U_0. Then

$$\widetilde{p}_{*x}(q_l([x])) = p_{*x}(e_l) = p_{*x}(\frac{\partial}{\partial z_l}) = \sum_{i=1}^{n} \frac{\partial(\zeta_i \circ p)}{\partial z_l}(x) \frac{\partial}{\partial \zeta_i}([x]) = \frac{1}{z_0} \frac{\partial}{\partial \zeta_l}([x]).$$

Moreover, $\beta(x) = (x, [x]) = z_0 \varepsilon_0([x])$. It follows

$$\varphi(q_l) = \frac{\partial}{\partial \zeta_l} \otimes \varepsilon_0.$$

Hence φ sends a frame of Q to a frame of $T^h\mathbb{P}(V) \otimes \mathcal{O}(-1)$, therefore φ is a vector bundle isomorphism. This implies $Q \otimes \mathcal{O}(1) \cong T^h\mathbb{P}(V) \otimes \mathcal{O}(-1) \otimes \mathcal{O}(1) \cong T^h\mathbb{P}(V) \otimes \{\text{the trivial bundle}\} \cong T^h\mathbb{P}(V)$. □

3.3 Complex Grassmann Manifolds

Let V be a complex vector space of dimension n and $0 < k < n$ a natural number. Denote the set of k-dimensional subspaces of V by $Gr(k, V)$. This set admits the structure of a complex manifold of dimension $k(n - k)$. To show this, let $A \in Gr(k, V)$ and let W be a $(n - k)$-dimensional subspace of V such that $A \cap W = \{0\}$, i.e., $V = A \oplus W$. Set

$$\Omega_{A, W} = \{E \in Gr(k, V) : E \cap W = \{0\}\}. \tag{3.5}$$

Clearly, the sets $\Omega_{A, W}$ cover $Gr(k, V)$. Every set $E \in \Omega_{A, W}$ can be uniquely represented as the graph $E = \{x + u_E(x) : x \in A\}$ of a linear map $u_E : A \to W$; this is the map $u_E(x) = -($ the component of x in W with respect to the decomposition $V = E \oplus W)$. In this way, we obtain a bijective map $\varphi_{A, W} : \Omega_{A, W} \to Hom(A, W)$ given by $E \to u_E$. Obviously, $u_E = 0$ exactly when $E = A$.

Now, let A' be another element of $Gr(k, V)$, and let W' be a $(n-k)$-dimensional subspace of V with $A' \cap W' = \{0\}$. Denote by $p_{A'} : A' \oplus W' \to A'$ the natural projection. If $u \in Hom(A, W)$, then $x + u(x) \in W'$ for some $x \in A$ exactly when $p_{A'}(x + u(x)) = 0$. Therefore $\varphi_{A,W}$ maps the intersection $\Omega_{A,W} \cap \Omega_{A',W'}$ onto the open set $\{u \in Hom(A, W) : p_{A'}(x+u(x)) \neq 0 \text{ for every } x \in A \setminus \{0\}\}$. Note that if u lies in this set, the map $p_{A'}|A + p_{A'} \circ u : A \to A'$ is injective, hence an isomorphism (since $dim\, A = dim\, A' = k$). Denote by $p_{W'} : A' \oplus W' \to W'$ the natural projection. Let $u \in Hom(A, W)$, and let $\varphi^{-1}_{A,W}(u) = graph\, v$, where $v \in Hom(A', W')$. Then for every $x \in A$ there is $y \in A'$ such that $x + u(x) = y + v(y)$. Clearly, $y = p_{A'}(x + u(x))$ and $v(y) = p_{W'}(x + u(x))$, hence $v \circ (p_{A'}|A + p_{A'} \circ u)(x) = (p_{W'}|A + p_{W'} \circ u)(x)$. It follows

$$\varphi_{A',W'} \circ \varphi^{-1}_{A,W}(u) = (p_{W'}|A + p_{W'} \circ u) \circ (p_{A'}|A + p_{A'} \circ u)^{-1} : A' \to W'. \quad (3.6)$$

This identity shows that $\varphi_{A',W'} \circ \varphi^{-1}_{A,W}$ is a rational holomorphic function of u.

Consider every vector space $Hom(A, W)$ with its natural structure of a complex manifold and transfer this structure on $\Omega_{A,W}$ by means of the bijective map $\varphi_{A,W}$. Identity (3.6) shows that the complex structures of $\Omega_{A,W}$ and $\Omega_{A',W'}$ agree on the intersection of these sets.

Thus, $Gr(k, V)$ becomes a complex manifold of dimension $k(n-k)$ which is called a **Grassmann manifold**.

Let $E \in \Omega_{A,W}$, and fix bases $\{a_1, \ldots, a_k\}$ of A and $\{a_{k+1}, \ldots, a_n\}$ of W. For $u = \varphi_{A,W}(E)$, set

$$u(a_i) = \sum_{r=k+1}^{n} z_{ir} a_r, \quad i = 1, \ldots, k. \quad (3.7)$$

Then $\{z_{ir}\}$ are local coordinates of E on the manifold $Gr(k, V)$.

Now, we are going to show that a sequence $\{E'\}$ of k-dimensional subspaces of V converges to a k-dimensional subspace with respect to the topology of $Gr(k, V)$ exactly when there exist a basis $\{e_1, \ldots, e_k\}$ of E and a basis $\{e'_1, \ldots, e'_k\}$ of E' such that for every $i = 1, \ldots, k$ the sequence $\{e'_i\}$ of vectors of V coverges to e_i.

If $\{E'\} \to E$ and $E \in \Omega_{A,W}$, we can assume that $E' \in \Omega_{A,W}$. Let $\{a_1, \ldots, a_k\}$ and $\{a_{k+1}, \ldots, a_n\}$ be bases of A and W. Set $u = \varphi_{A,W}(E)$ and $u' = \varphi_{A,W}(E')$. Then $\{u'\} \to u$, hence $\{a_i + u'_i(a_i)\} \to a_i + u(a_i)$ for every $i = 1, \ldots, k$. The vectors $e'_i = a_i + u'_i(a_i)$ constitute a basis of E' and the vectors $e_i = a_i + u(a_i)$ form a basis of E.

Conversely, suppose $e = \{e_1, \ldots, e_k\}$ and $e' = \{e'_1, \ldots, e'_k\}$ are bases of E and E' such that $\{e'_i\} \to \{e_i\}$ for every $i = 1, \ldots, k$. The space E lies in a set $\Omega_{A,W}$, and let $\{a_1, \ldots, a_k\}$ and $\{a_{k+1}, \ldots, a_n\}$ be bases of A and W. Set $e_i = \sum_{p=1}^{n} \alpha_{ip} a_p$, $e'_i = \sum_{p=1}^{n} \alpha'_{ip} a_p$. The n-tuple $\{e_1, \ldots, e_k, a_{k+1}, \ldots, a_n\}$ is a basis of V, and the determinant of the transition matrix from this basis to the basis $\{a_1, \ldots, a_k, a_{k+1}, \ldots, a_n\}$ equals $det\,[\alpha_{\nu\mu}]^k_{\nu,\mu=1}$, so $det\,[\alpha_{\nu\mu}]^k_{\nu,\mu=1} \neq 0$. The sequence $\{\alpha'_{ij}\} \to \alpha_{ij}$, hence $det\,[\alpha'_{\nu\mu}]^k_{\nu,\mu=1} \neq 0$ if e' is sufficiently close to e. Then $\{e'_1, \ldots, e'_k, a_{k+1}, \ldots, a_n\}$ is a basis of V. This means that $E' \cap W = \{0\}$, thus $E' \in \Omega_{A,W}$. Set $u = \varphi_{A,W}(E)$ and $u(a_i) = \sum_{r=k+1}^{n} z_{ir} a_r$, $i = 1, \ldots, k$. The set of vectors $\{a_1+u(a_1), \ldots, a_k+u(a_k)\}$ is a basis of E, hence $e_i = \sum_{j=1}^{k} \lambda_{ij}(a_j + u(a_j)) = \sum_{j=1}^{k} \lambda_{ij}a_j + \sum_{r=k+1}^{n} \sum_{j=1}^{k} \lambda_{ij}z_{jr}a_r$, where $\lambda_{ij} \in \mathbb{C}$. It follows $\lambda_{ij} = \alpha_{ij}$ for $i, j = 1, \ldots, k$ and $\sum_{j=1}^{k} \lambda_{ij}z_{jr} = \alpha_{ir}$ for $r = k+1, \ldots, n$. Therefore if the entries of the matrix $[z_{ir}]$ are the local coordinates of E, $[z_{jr}] = [\alpha_{ij}]^{-1}[\alpha_{ir}]$. Similarly, $[z'_{jr}] = [\alpha'_{ij}]^{-1}[\alpha'_{ir}]$ are the local coordinates of E'. Clearly, $\{[z'_{jr}]\} \to [z_{jr}]$, hence $\{E'\} \to E$.

One often uses on $Gr(k, V)$ a cover of coordinate neighbourhoods defined in the following way: Let $v_1, \ldots, v_n$ be a basis of V. For every set $\alpha = (\alpha_1, \ldots, \alpha_k)$ of natural numbers with $1 \le \alpha_1 < \cdots < \alpha_k \le n$, denote by $\overline{\alpha} = (\overline{\alpha}_1, \ldots, \overline{\alpha}_{n-k}), \overline{\alpha}_1 < \cdots < \overline{\alpha}_{n-k}$ the set that completes α to $\{1, 2, \ldots, n\}$. Put $A_\alpha = span\{v_{\alpha_1}, \ldots, v_{\alpha_k}\}$, $A_{\overline{\alpha}} = span\{v_{\overline{\alpha}_1}, \ldots, v_{\overline{\alpha}_{n-k}}\}$ Then the coordinate neighbourhoods $\Omega_{A_\alpha, A_{\overline{\alpha}}}$ cover $Gr(k, V)$. Indeed, let $E \in Gr(k, V)$ and let $e_i = \sum_{p=1}^{n} \lambda_{ip}v_p, i = 1, \ldots, k$, be a basis of E. Then the $(k \times n)$-matrix $[\lambda_{ip}]$ is of rank k. Let $[\lambda_{i\,\alpha_j}]$, $1 \le \alpha_1 < \cdots < \alpha_k \le n$, be a non-singular minor. If $\sum_{i=1}^{k} \xi_i e_i = \sum_{i=1}^{k} \sum_{p=1}^{n} \xi_i \lambda_{ip} v_p$ is a vector of E lying in $A_{\overline{\alpha}}$, then $\sum_{i=1}^{k} \xi_i \lambda_{ip} = 0$ for $p = \alpha_1, \ldots, \alpha_k$, hence $\xi_i = 0, i = 1, \ldots, k$. Therefore $E \cap A_{\overline{\alpha}} = \{0\}$, thus $E \in \Omega_{A_\alpha, A_{\overline{\alpha}}}$.

Problem 3.3.1 Prove that the atlas on the projective space defined in Sect. 3.1 of this chapter and the atlas $\{\Omega_{A,W}, \varphi_{A,W}\}$ on $Gr(1, V) = \mathbb{P}(V)$ determine the same complex structure.

Problem 3.3.2 Let h be a Hermitian metric on V. Then the map that assigns to $E \in Gr(k, V)$ its orthogonal complement $E^\perp$ with respect to h is a diffeomorphism of $Gr(k, V)$ onto $Gr(n-k, V)$, which is not a holomorphic map.

Problem 3.3.3 Let W be a complex subspace of V. Prove that $Gr(k, W)$ is an embedded complex submanifold of $Gr(k, V)$.

Proposition 3.3.1 *The Grassmann manifold $Gr(k, V)$ is Hausdorff, connected, and compact. Thus, second-countable and paracompact.*

Proof Let $E_1, E_2 \in Gr(k, V)$. Then there is a $(n-k)$-dimensional subspace W of V such $E_1 \cap W = \{0\}$ and $E_2 \cap W = \{0\}$. For, take a basis $a_1, \ldots, a_r$ of the space $E_1 \cap E_2$ and extend this basis to a basis $a_1, \ldots, a_r, e'_1, \ldots, e'_{k-r}$ of E_1 and a basis $a_1, \ldots, a_r, e''_1, \ldots, e''_{k-r}$ of E_2. Set $u_i = e'_i + e''_i$, $i = 1, \ldots k-r$, and $U = span\{u_1, \ldots, u_{n-r}\}$. Then $E_1 + E_2 = E_1 \oplus U = E_2 \oplus U$. If $E_1 + E_2 = V$, set $W = U$. If $E_1 + E_2 \neq V$, take a complement U' of $E_1 + E_2$ in V, $V = (E_1 + E_2) \oplus U'$, and set $W = U + U'$.

Choosing a space W as above, we have $E_2 \in \Omega_{E_1, W}$. The set $\Omega_{E_1, W}$ is homeomorphic to an open subset of $\mathbb{C}^{k(n-k)}$. Hence if $E_1 \neq E_2$, then there are disjoint neighbourhoods of E_1 and E_2.

Fix a Hermitian metric h on V, and denote by $U(V)$ the group of linear transformations of V preserving h. This group acts transitively on $Gr(k, V)$. To see this, if $E', E'' \in Gr(k, V)$, take h-orthonormal bases $e'_1, \ldots, e'_k$ and $e''_1, \ldots, e''_k$ of E' and E''. Complete these bases to h-orthonormal bases $e'_1, \ldots, e'_n$ and $e''_1, \ldots, e''_n$ of the space V. Then the linear transformation of V that maps e'_i to e''_i, $i = 1, \ldots, n$, preserves the metric h and sends E' onto E''. Now, let E_0 be a (fixed) k-dimensional subspace of V. Because of the transitive action of $U(V)$, the Grassmannian $Gr(k, V)$ is the image of $U(V)$ under the map $U(V) \ni T \to T(E_0) \in Gr(k, V)$. This map is continuous and the group $U(V) \cong U(n)$ is connected and compact. Hence so is the Grassmann manifold. □

Under the notation of the proof above, if $E_0 \in Gr(k, V)$, then the subgroup of $U(V)$ that fixes E_0 is $U(E_0) \times U(E_0^{\perp})$. The quotient group $U(V)/(U(E_0) \times U(E_0^{\perp}))$ has the structure of a smooth manifold and, with this structure, it is diffeomorphic to the Grassmann manifold $Gr(k, V)$. This is a consequence of a general result in Lie group theory (see, e.g., Brickell and Clark [3], Chap. 13, Proposition 13.3.3) some facts of which will be given in the next section.

Considerations similar to that above can also be made in the case when V is a real vector space. In this case, the set of k-dimensional subspaces of V admits the structure of a real-analytic manifold. In order to see that the real Grassmann manifold is compact, we can use the group $O(V)$ of orthogonal transformations with respect to a metric on V. However, this group is not connected and to prove that the Grassmann manifold is connected, we should fix a metric and, in addition, an orientation on V. Then we take the group $SO(V)$ of orthogonal transformations that preserve the orientation instead of $O(V)$. In the real case, if $E_0 \in Gr(k, V)$, then $Gr(k, V) \cong SO(V)/(SO(E_0) \times SO(E_0^{\perp}))$.

3.4 The Tautological (Universal) Bundle over a Complex Grassmann Manifold

Let $\underline{V} = Gr(k, V) \times V$ be the trivial bundle over the Grassmann manifold, and let

$$S = \{(E, v) \in Gr(k, V) \times V : v \in E\}.$$

The set S admits a natural structure of a vector subbundle of the trivial bundle $\underline{V}$. Indeed, in the notation of the preceding section, let $\Omega_{A,W}$ be a coordinate neighbourhood of the Grassmann manifold. Take bases $\{a_1, \ldots, a_k\}$ of A and $\{a_{k+1}, \ldots, a_n\}$ of W. Define sections of $\underline{V}$ by $\sigma_i(E) = (E, a_i + u_E(a_i))$, $i = 1, \ldots, k$, $\sigma_j(E) = (E, a_j)$, $j = k+1, \ldots, n$ where $u_E : A \to W$ is the linear map whose graph is E. Obviously, these sections constitute a frame of $\underline{V}$, and $\sigma_i(E) \in E$ for $i = 1, \ldots, k$. Then, by Chap. 1, Proposition 1.7.3, S admits the structure of a subbundle of $\underline{V}$ for which $\sigma_1, \ldots, \sigma_k$ is a frame of the subbundle S. The fibre of the bundle S over a point E of $Gr(k, V)$ is the k-dimensional subspace E of V itself. Therefore the name **tautological** for the bundle S. It is also called

universal, because it possesses the important property that every vector bundle is a pull-back of the tautological bundle.

Proposition 3.4.1 *Let M be a paracompact manifold of dimension n, and let s, k be natural numbers with $s > k + n$. Then for every (smooth) complex vector bundle $E \to M$ of rank k over M there is a smooth map $f_E : M \to Gr(k, \mathbb{C}^s)$ such that $f_E^* S \cong E$ where $S \to Gr(k, \mathbb{C}^s)$ is the tautological bundle. Moreover, a map f_E with this property is unique up to homotopy. A complex vector bundle $E' \to M$ of rank k is isomorphic to E if and only if the maps f_E and $f_{E'}$ are homotopic.*

A proof of this statement can be seen in Hirsch [11], Chap. 4, Sect. 3, Theorem 3.4. Note that a similar claim for holomorphic vector bundles is not true (see the comments after Theorem 2.17 in Wells [25], Chap. I, Sect. 2).

If $Q = \underline{V}/S$ is the quotient bundle, the sequence of vector bundles over the Grassmann manifold $Gr(k, V)$

$$0 \longrightarrow S \overset{inclusion}{\longrightarrow} \underline{V} \overset{projection}{\longrightarrow} Q \longrightarrow 0$$

is exact.

Remark 3.4.1 In Bott and Tu [2], a result (Proposition 23.9) similar to the first part of Proposition 3.4.1 is proved, but the Grassmannian $Gr(k - r, \mathbb{C}^k)$, denoted in the book by $G_r(\mathbb{C}^k)$, and the quotient bundle $Q = Q_r$ over it are used instead of the Grassmann manifold $Gr(r, \mathbb{C}^k)$ and the universal bundle S. As it has been stated in Problem 3.3.2, the manifold $G_r(\mathbb{C}^k) = Gr(k - r, \mathbb{C}^k)$ is diffeomorphic to $Gr(r, \mathbb{C}^k)$ via the map given by $E \to E^\perp$, where $E^\perp$ is the orthogonal complement of E with respect to the standard Hermitian metric of $\mathbb{C}^k$. If $E \in Gr(k - r, \mathbb{C}^k)$, the fibre of Q_r over E is $(Q_r)_E = \mathbb{C}^k/E \cong E^\perp = S_{E^\perp}$, where $S_{E^\perp}$ is the fibre of the universal bundle $S \to Gr(r, \mathbb{C}^k)$. The family of fibre isomorphisms $(Q_r)_E \to S_{E^\perp}$ defines a diffeomorphism $Q_r \to S$ covering the diffeomorphism $Gr(k-r, \mathbb{C}^k) \to Gr(r, \mathbb{C}^k)$. Therefore the statement in Bott and Tu [2] quoted above is equivalent to the first part of Proposition 3.4.1.

In the particular case $k = 1$, $Gr(1, V) = \mathbb{P}(V)$ and $S = \mathcal{O}(-1)$.

The case $k = n - 1$ ($n = dim\ V$) is dual in the following sense. Let $f : Gr(n - 1, V) \to \mathbb{P}(V^*)$ be the biholomorphic map that maps a (complex) hyperplane $E \subset V$ to the line L_E in V^* of linear forms $\xi : V \to \mathbb{C}$ vanishing on E; the inverse map sends $[\xi] \in \mathbb{P}(V^*)$ to the hyperplane $E = Ker\, \xi$. Let $\underline{V}^*$, $\mathcal{O}_{V^*}(-1)$ and $\mathcal{O}_{V^*}(1)$ be, respectively, the trivial and tautological bundles, and the bundle dual to the tautological one over $\mathbb{P}(V^*)$. Set $Q_{V^*} = \underline{V}^*/\mathcal{O}_{V^*}(-1)$. Then we can define biholomorphic maps $\widetilde{f} : S \to (Q_{V^*})^*$ and $\widehat{f} : Q \to \mathcal{O}(1)_{V^*}$ that are fibre preserving and isomorphisms on the fibres. For these maps, the diagrammes

$$\begin{array}{ccc} S & \xrightarrow{\widetilde{f}} & (Q_{V^*})^* \\ \downarrow & & \downarrow \\ Gr(n-1, V) & \xrightarrow[f]{} & \mathbb{P}(V^*) \end{array}$$

and

$$\begin{array}{ccc} Q & \xrightarrow{\widehat{f}} & \mathcal{O}_{V^*}(1) \\ \downarrow & & \downarrow g \\ Gr(n-1, V) & \xrightarrow[f]{} & \mathbb{P}(V^*) \end{array}$$

are commutative. Thus, $S \cong f^*(Q_{V^*})^*$ and $Q \cong f^*\mathcal{O}_{V^*}(1)$. The diffeomorphisms $\widetilde{f}$ and $\widehat{f}$ are defined as follows. The map $\widetilde{f}$ assigns to $(E, v) \in S$ the linear form $\widetilde{\omega} : (Q_{V^*})_{f(E)} = V^*/L_E \to \mathbb{C}$ defined by $\widetilde{\omega}([\alpha]) = \alpha(v)$, where $[\alpha]$ is the equivalence class of $\alpha \in V^*$. Also, the map $\widehat{f}$ assigns to $[(E, v)] \in Q$ the linear form $\widehat{\omega} : (\mathcal{O}_{V^*}(1))_{f(E)} = (L_E)^* \to \mathbb{C}$ defined by $\widehat{\omega}(\xi) = \xi(v)$.

Problem 3.4.1 Check that the maps f, $\widetilde{f}$ and $\widehat{f}$ possess the properties claimed above.

To prove the next statement, we need some basic facts about Lie groups, see Brickell and Clark [3], Chaps. 12 and 13, Helgason [10], Chap. II, Warner [24], Chap. 3. Recall that, by definition, a Lie group is a group that possesses the structure of a smooth manifold such that the group operation $(g_1, g_2) \to g_1 g_2$ is a smooth map $G \times G \to G$. In this case, the map $G \ni g \to g^{-1} \in G$ is also smooth. If H is a closed subgroup of a Lie group G, then H admits a unique structure of an embedded submanifold of G such that H is a Lie group (the smooth structure of H can be null dimensional, i.e., H can be a discrete subset of G). For a closed subgroup H of G, the quotient space G/H of left (right) cosets admits a unique structure of a (Hausdorff) manifold for which the natural projection $\mu : G \to G/H$ is a submersion. Since μ is a submersion, $H = \mu^{-1}([e])$ is an embedded submanifold of dimension $dim\, G - dim\, G/H$, so $dim\, G/H = dim\, G - dim\, H$. Moreover, $T_e H = Ker\, \mu_{*e}$, hence μ_* yields an isomorphism $\widetilde{\mu} : T_e G / T_e H \to T_{[e]}(G/H)$ given by $\widetilde{\mu}([X]) = \mu_{*e}(X)$. Note also that the natural action of G on G/H is smooth, i.e., the map $G \times G/H \to G/H$ defined by $(f, [g]) \to [fg]$ is smooth.

Now, suppose that a Lie group G acts smoothly on a manifold M. Fix a point m of M, and denote the stationary (isotropy) subgroup of G at m by H_m, $H_m = \{g \in G : g(m) = m\}$. Then H_m is a closed subgroup of G, hence G/H_m is a smooth manifold. This manifold is diffeomorphic to M provided the action of G is transitive and G has a countable base of its topology (the latter holds if the group G is connected); the diffeomorphism between G/H_m and M is given by $[g] \to g(m)$.

Besides smooth groups, we can consider complex Lie groups. These are groups G with the structure of a complex manifold such that the group operation is a holomorphic map $G \times G \to G$. Unlike smooth Lie groups, not every closed subgroup H of a complex Lie group G is a complex Lie group. For example, $G = GL(\mathbb{C}^n)$ is a complex Lie group and the unitary $H = U(n)$ is a compact connected subgroup of G. The group $U(n)$ does not admit any structure of a complex submanifold of $GL(\mathbb{C}^n)$, otherwise the coordinate functions $[a_{ij}] \to a_{ij}$ would be holomorphic on the compact connected complex manifold $U(n)$, hence constant on $U(n)$ by the maximum modulus principle.

If G is a complex Lie group and H is a closed complex subgroup, the quotient manifold G/H admits a natural structure of a complex manifold. Indeed, let $J : TG \to TG$ be the complex structure of G. Since H is a complex submanifold of G, the tangent spaces of H (considered as subspaces of TG) are invariant under J. Hence J gives rise to a linear operator $\widehat{J} : T_eG/T_eH \to T_eG/T_eH$ defined by $\widehat{J}[X] = [JX]$. As noted above, if $\mu : G \to G/H$ is the natural projection, the differential μ_* yields an isomorphism $\widetilde{\mu} : T_eG/T_eH \to T_{[e]}(G/H)$. Setting $\widetilde{J} = \widetilde{\mu} \circ \widehat{J} \circ \widetilde{\mu}^{-1}$, we get a linear operator $\widetilde{J} : T_{[e]}(G/H) \to T_{[e]}(G/H)$ with $\widetilde{J}^2 = -Id$. For every $g \in G$, denote by $L_g : G/H \to G/H$ the natural action of g on G/H. Then $\widetilde{J}_{[g]} = (L_g)_* \circ \widetilde{J} \circ (L_{g^{-1}})_*$ is a well-defined almost complex structure on G/H invariant under the action of G. It can be shown that the Nijenhuis tensor of $\widetilde{J}$ vanishes (see Kobayashi and Nomizu [14], vol. II, Chap. 10, Sect. 6, Example 6.2). Therefore G/H admits the structure of a complex manifold with complex structure $\widetilde{J}$. It is clear that $\mu_* \circ J = \widetilde{J} \circ \mu_*$, i.e., μ is a holomorphic map.

Next, let G be a complex Lie group acting holomorphically on a complex manifold M, i.e., the map $\psi : G \times M \to M$ defined by $\psi(g, x) = g(x)$ is holomorphic. Fix a point m of M and suppose that the stationary subgroup H_m of G at the point m is a complex subgroup of G. Suppose further that G acts transitively and that G has a countable base of its topology. Let $\alpha : G/H_m \to M$ be the diffeomorphism defined by $\alpha([g]) = g(m)$. We claim that α is a biholomorphism. To prove this, let $\psi_m : G \to M$ be the map $\psi_m(g) = g(m)$. This map is holomorphic by the assumption that the action of G is holomorphic. Since the natural projection $\mu : G \to G/H_m$ is a holomorphic submersion, around every point of G, there are complex coordinates in which μ is given by $(\zeta_1, \dots, \zeta_k) \to (\zeta_1, \dots, \zeta_l, 0, \dots, 0) \in \mathbb{C}^k$, $l = dimG/H_m$. This implies that in a neighbourhood of every point of G/H_m, there is a holomorphic map $s : G/H_m \to G$ such that $\mu \circ s = Id$ (see, if necessary, Brickell and Clark [3], Chap. 6, Proposition 6.1.4). Clearly, $\alpha = \psi_m \circ s$, hence α is a holomorphic map. Since α is diffeomorphism, the inverse map α^{-1} is also holomorphic, i.e., the map α is biholomorphic. In conclusion, the complex manifold M is not only diffeomorphic to G/H_m, but it is biholomorphic to this quotient manifold as well.

Problem 3.4.2 Show that the smooth map α^{-1} is holomorphic.

A theorem due to G. R. Clements (1912) states that every bijective holomorphic map between two domains in $\mathbb{C}^n$ is biholomorphic, i.e., the inverse map is also holomorphic; for a proof, see Griffiths and Harris [9], Chap. 0, Sect. 2, Sec. Submanifolds and Subvarieties. Note that this theorem is not true for smooth maps even in dimension 1, take the function $f(t) = t^3$.

Problem 3.4.3 Prove that the complex Lie group $G = GL(V)$ acts holomorphically on the Grassmannian $Gr(k, V)$.

Problem 3.4.4 Show that the stationary subgroup H_A of G at a point $A \in Gr(k, V)$ is a closed complex subgroup of G.

Proposition 3.4.2 *The holomorphic tangent bundle $T^h Gr(k, V)$ of the Grassmann manifold is canonically isomorphic to $Hom(S, Q) \cong .S^* \otimes Q$.*

Proof The complex linear group $G = GL(V)$ acts holomorphically and transitively on $Gr(k, V)$. Let $A \in Gr(k, V)$. The stationary subgroup H_A of G at the point A is a closed complex subgroup of G. Hence the map $\alpha_A : G/H_A \to Gr(k, V)$, $[g] \to g(A)$, is biholomorphic. Then $(\alpha_A^{-1})_{*A} : T_A^h Gr(k, V) \to T_{[I]}^h(G/H_A)$ is a vector space isomorphism; here and hereafter I stands for the identity of V. Let $\mu_A : G \to G/H_A$ be the natural projection, and $\widetilde{\mu}_A : T_I^h G/T_I^h H_A \to T_{[I]}^h(G/H_A)$ the isomorphism induced by $(\mu_A)_{*I}$. Then $\widetilde{\mu}_A^{-1} \circ (\alpha_A^{-1})_{*A}$ is an isomorphism of $T_A^h Gr(k, V)$ onto $T_I^h G/T_I^h H_A$. The group $G = GL(V)$ is an open subset of the vector space $gl(V) = Hom(V, V)$. If $\kappa_I : T_I^h gl(V) \to gl(V)$ is the canonical isomorphism, $\kappa_I(T_I^h G) = gl(V)$ and $\kappa_I(T_I^h H_A) = \{L \in gl(V) : L(A) \subset A\}$. Denoting the latter vector space by $gl_A(V)$, we have an isomorphism $\widetilde{\kappa}_A : T_I^h G/T_I^h H_A \to gl(V)/gl_A(V)$ defined by $\widetilde{\kappa}_A([X]) = [\kappa_I(X)]$. Let $p_A : V \to V/A$ be the natural projection. Then the space $gl(V)/gl_A(V)$ is isomorphic to the space $Hom(V, V/A)/\{\widetilde{L} : \widetilde{L}(A) = 0\}$ via the map $[L] \to [p_A \circ L]$. The latter space is isomorphic to $Hom(A, V/A) = Hom(S_A, Q_A)$ by the map $[\widetilde{L}] \to \widetilde{L}|A$. Thus, we obtain an isomorphism $\chi_A : gl(V)/gl_A(V) \to Hom(S_A, Q_A)$ defined by $\chi_A([L]) = p_A \circ L|A$. Now, set $F_A = \chi_A \circ \widetilde{\kappa}_A \circ \widetilde{\mu}_A^{-1} \circ (\alpha_A^{-1})_{*A}$. The maps F_A yield a fibre-preserving map F of the bundle $T^h Gr(k, V)$ onto the bundle $Hom(S, Q)$. We shall show that this map is an isomorphism.

Let $\Omega_{A,W}$ be a standard coordinate neighbourhood of A. Take a basis $a = (a_1, \ldots, a_n)$ of V such that $a_1, \ldots, a_k$ is a basis of A and $a_{k+1}, \ldots, a_n$ a basis of W. Using this basis of V, we define local coordinates $\{z_{ir}\}$, $1 \le i \le k$, $k+1 \le r \le n$, of $Gr(k, V)$ on $\Omega_{A,W}$ as in (3.7). Clearly, $z_{ir}(A) = 0$, and for every $E \in \Omega_{A,W}$, the vectors $b_i(E) = a_i + \sum_{r=k+1}^{n} z_{ir}(E) a_r$, $i = 1, \ldots, k$, constitute a basis of E. Set $b_r(E) = a_r$ for $r = k+1, \ldots, n$. Then $b(E) = (b_1(E), \ldots, b_n(E))$ is a basis of V. Fix an arbitrary point $E \in \Omega_{A,W}$. For every $P \in \Omega_{A,W}$, let g_P be the linear transformation of V that sends the basis $b(E)$ to the basis $b(P)$. Denote the map $P \to g_P$ by $\beta : \Omega_{A,W} \to G = GL(V)$. Then $\alpha_E^{-1} = \mu_E \circ \beta$. If $pr_E : T_I G \to T_I G/T_I H_E$ is the natural projection, $\widetilde{\mu}_E \circ pr_E = (\mu_E)_{*I}$. Therefore $(\alpha_E^{-1})_{*E} = (\mu_E)_{*I} \circ \beta_{*E} = \widetilde{\mu}_E \circ pr_E \circ \beta_{*E}$. Hence $\widetilde{\mu}_E^{-1} \circ (\alpha_E^{-1})_{*E} = pr_E \circ \beta_{*E}$.

Denote the coordinates of the point E by $\{z_{ir}^0\}$, and set $b_l^0 = b_l(E), l = 1, \dots, n$. Then for every point $P \in \Omega_{A,W}$ with coordinates $\{z_{ir}\}$,

$$g_P(a_i) = g_P(b_i^0 - \sum_{r=k+1}^{n} z_{ir}^0 b_r^0) = b_i(P) - \sum_{r=k+1}^{n} z_{ir}^0 b_r(P) = a_i + \sum_{r=k+1}^{n} (z_{ir} - z_{ir}^0) a_r.$$

Let $\{\zeta_{pq}\}_{p,q=1}^n$ be the standard coordinates of $gl(V)$ defined by the basis $a_1, \dots, a_n$ of V. So if $L \in gl(V)$, then $\zeta_{pq}(L)$ are the entries of the matrix of the linear map L with respect to the basis $a_1, \dots, a_n$. It is obvious that

$$\zeta_{pq} \circ \beta = \begin{cases} \delta_{pq} & \text{if } 1 \le p, q \le k \quad \text{or} \quad k+1 \le p, q \le n \\ z_{pq} - z_{pq}^0 & \text{if } 1 \le p \le k \quad \text{and} \quad k+1 \le q \le n \\ 0 & \text{if } k+1 \le p \le n \quad \text{and} \quad 1 \le q \le k. \end{cases}$$

Therefore β is a holomorphic map and

$$\beta_*(\frac{\partial}{\partial z_{ir}}(E)) = \sum_{p,q=1}^{n} \frac{\partial(\zeta_{pq} \circ \beta)}{\partial z_{ir}}(E) \frac{\partial}{\partial \zeta_{pq}}(I) = \frac{\partial}{\partial \zeta_{ir}}(I).$$

Thus,

$$\widetilde{\mu}_E^{-1} \circ (\alpha_E^{-1})_{*E}(\frac{\partial}{\partial z_{ir}}(E)) = [\frac{\partial}{\partial \zeta_{ir}}(I)] \in T_I^h G / T_I^h H_E.$$

The isomorphism κ_I sends $\dfrac{\partial}{\partial \zeta_{ir}}(I)$ to the element $L_{ir} \in gl(V)$ with coordinates ζ_{ir}, i.e., to the endomorphism L_{ir} of V given by $L_{ir}(a_i) = a_r$, $L_{ir}(a_l) = 0$ for $l \neq i$, $1 \le l \le n$. Hence

$$\widetilde{\kappa}_E \circ \widetilde{\mu}_E^{-1} \circ (\alpha_E^{-1})_{*E}(\frac{\partial}{\partial z_{ir}}(E)) = [L_{ir}] \in gl(V)/gl_E(V).$$

It follows that the map F sends $\dfrac{\partial}{\partial z_{ir}}(E)$ to the linear map $Z_{ir} : E \to V/E$ for which $Z_{ir}(a_i) = p_E(a_r)$ and $Z_{ir}(a_l) = 0$ for $l \neq i$, $1 \le l \le n$.

The sections $\sigma_j : \Omega_{A,W} \ni E \to b_j(E)$, $1 \le j \le k$, of S form a (holomorphic) frame, and the sections $\sigma_s : E \to p_E(a_s)$, $k+1 \le s \le n$, a frame of Q on $\Omega_{A,W}$. Denote by σ_{js} the section of $Hom(S, Q)$ that at every point of $\Omega_{A,W}$ is given by $\sigma_j \to \sigma_s$ and $\sigma_l \to 0$ for $l \neq j$, $1 \le l \le n$. Clearly, the sections σ_{js} form a holomorphic frame of $Hom(S, Q)$ over $\Omega_{A,W}$. By the considerations above, F maps the frame $\{\dfrac{\partial}{\partial z_{ir}}\}$ of $T^h Gr(k, V)$ to the frame $\{\sigma_{ir}\}$ of $Hom(S, Q)$. Hence F is a bundle isomorphism. □

Let $p : \underline{V} \to Q = \underline{V}/S$ be the natural projection. The map p induces a linear map $p^0 : H^0(Gr(k, V), \underline{V}) = V \to H^0(Gr(k, V), Q)$ of the spaces of global holomorphic sections defined by $p^0(s) = p \circ s$.

Proposition 3.4.3 *The linear map* $p^0 : H^0(Gr(k, V), \underline{V}) \to H^0(Gr(k, V), Q)$ *is an isomorphism.*

Proof Let $v_1, \ldots, v_n$ be a basis of V and let $s_1, \ldots, s_n$ be the sections of $\underline{V}$ yielded by this basis, i.e., defined by $s_i(E) = (E, v_i), i = 1, \ldots, n$. These sections constitute a basis of the space $H^0(Gr(k, V), \underline{V})$. Their images under p^0, the sections $p^0(s_i)$ of Q, are linearly independent over $\mathbb{C}$. Indeed, suppose $\sum_{i=1}^n \lambda_i p^0(s_i) = 0$ for some constants $\lambda_i \in \mathbb{C}$. Then for every $E \in Gr(k, V)$, we have $\sum_{i=1}^n \lambda_i v_i \in E$ (recall that $Q_E = V/E$). Hence if $E = span\{v_{i_1}, \ldots, v_{i_k}\}$ for some set of different indices $i_1, \ldots, i_k$, then $\lambda_i = 0$ for $i \neq i_1, \ldots, i_k$. It follows that $\lambda_i = 0$ for every $i = 1, ..., n$ Now, we shall show that the linearly independent sections $p^0(s_i)$ generate the space $H^0(Gr(k, V), Q)$.

Let $E_0 \in Gr(k, V)$ and take a complex subspace W of V such that $E_0 \subset W$ and $dim\, W = k + 1$. Also, take a subspace W' of V for which $W \oplus W' = V$. The Grassmannian $Gr(k, W)$ is an embedded complex submanifold of $Gr(k, V)$. As it has been observed above, $Gr(k, W) \cong \mathbb{P}(W^*)$ by the biholomorphism f that sends $E \in Gr(k, W)$ to the projective class $[\xi]$ of a 1-form ξ on W such that $E = Ker\, \xi$. Let $\mathcal{O}(-1)$ is the tautological bundle over $\mathbb{P}(W^*)$. Then $W/E \cong (\mathbb{C}.\xi)^* = \mathcal{O}(1)_{[\xi]}$ by the isomorphism that assigns to $[w] \in W/E$ the 1-form ω on $\mathbb{C}.\xi$ defined by $\omega(\xi) = \xi(w)$. Set $\overline{Q} = Q|Gr(k, W)$. Then for $E \in Gr(k, W)$,

$$\overline{Q}_E = (W \oplus W')/E \cong (W/E) \oplus W' \cong \mathcal{O}(1)_{[\xi]} \oplus W'. \tag{3.8}$$

Let $\underline{W}' = \mathbb{P}(W^*) \times W'$. Denote by $F : \overline{Q} \to \mathcal{O}(1) \oplus \underline{W}'$ the map that is the isomorphism (3.8) on every fibre. This map is biholomorphic (prove it). In this way, we obtain the following commutative diagramme:

$$\begin{array}{ccc} \overline{Q} & \xrightarrow{F} & \mathcal{O}(1) \oplus \underline{W}', \\ \downarrow & & \downarrow \\ Gr(k, W) & \xrightarrow[f]{} & \mathbb{P}(W^*) \end{array}$$

where f is a biholomorphic map and F is a bundle isomorphism. This diagramme induces an isomorphism in an obvious way:

$$H^0(Gr(k, W), \overline{Q}) \cong H^0(\mathbb{P}(W^*), \mathcal{O}(1) \oplus \underline{W}'). \tag{3.9}$$

By Proposition 3.2.1,

$$H^0(\mathbb{P}(W^*), \mathcal{O}(1)) \cong W. \tag{3.10}$$

This isomorphism assigns to every holomorphic section τ of $\mathcal{O}(1)$ the vector $u \in W$ defined by the condition that for every $\xi \in W^* \setminus \{0\}$ the identity $\xi(u) = \tau([\xi])(\xi)$ holds. It follows from (3.10) that

$$H^0(\mathbb{P}(W^*), \mathcal{O}(1) \oplus \underline{W}') \cong H^0(\mathbb{P}(W^*), \mathcal{O}(1)) \oplus H^0(\mathbb{P}(W^*), \underline{W}') \cong W \oplus W'.$$

Now, taking into account (3.9), we get

$$H^0(Gr(k, W), \overline{Q}) \cong W \oplus W'. \tag{3.11}$$

For brevity, set $\sigma_i = p^0(s_i)$, $i = 1, \ldots, n$. The restriction $\overline{\sigma}_i = \sigma_i|Gr(k, W)$ is a holomorphic section of $\overline{Q}$. Denote by (τ_i, z_i) the section of $\mathcal{O}(1) \oplus \underline{W}'$ that corresponds to $\overline{\sigma}_i$ under the isomorphism (3.9). Let $\xi \in W^* \setminus \{0\}$ and $E = Ker\,\xi$. Then $\overline{\sigma}_i(E) = (E, [v_i]_E)$, where $[v_i]_E$ is the coset of v_i in the quotient-space V/E. Write v_i as $v_i = w_i + w_i'$ with $w_i \in W$, $w_i' \in W'$. Let ω_i be the linear form on $\mathcal{O}(1)_{[\xi]} = \mathbb{C}.\xi$ defined by $\omega_i(\xi) = \xi(w_i)$. Under the isomorphism $F : \overline{Q} \to \mathcal{O}(1) \oplus \underline{W}'$, $\overline{\sigma}_i(E)$ goes to the element $\omega_i + ([\xi], w_i')$ of $\mathcal{O}(1)_{[\xi]} \oplus \underline{W}'_{[\xi]}$. Hence $(\tau_i([\xi]), z_i([\xi])) = (\omega_i, ([\xi], w_i'))$. Let u_i be the vector in W corresponding to τ_i under the isomorphism (3.10). Then $\xi(w_i) = \omega_i(\xi) = \tau_i([\xi])(\xi) = \xi(u_i)$. The identity $\xi(w_i) = \xi(u_i)$ holds for every $\xi \in W^* \setminus \{0\}$, therefore $w_i = u_i$. Thus, the section $\overline{\sigma}_i$ goes to $w_i + w_i' = v_i$ under the isomorphism (3.11). Therefore the holomorphic sections $\overline{\sigma}_i$, $i = 1, \ldots, n$, constitute a basis of $H^0(Gr(k, W), \overline{Q})$.

Now, let σ be a holomorphic section of Q. Set $\overline{\sigma} = \sigma|Gr(k, W)$. Then $\overline{\sigma} = \sum_{i=1}^n \mu_i \overline{\sigma}_i$ for some constants $\mu_i \in \mathbb{C}$. In particular, $\sigma(E_0) = \sum_{i=1}^n \mu_i \sigma_i(E_0)$. This shows that, at every point of $Gr(k, V)$, σ is a linear combination $\sigma = \sum_{i=1}^n \varphi_i \sigma_i$. The coefficients φ_i are holomorphic functions on $Gr(k, V)$. The Grassmannian is a compact and connected complex manifold, hence these functions are constant.

With this, we show that the sections $\sigma_i = p^0(s_i)$, $i = 1, \ldots, n$, form a basis of the vector space $H^0(Gr(k, V), Q)$, and the proposition is proved. □

Problem 3.4.5 Prove that the map $F : \overline{Q} \to \mathcal{O}(1) \oplus \underline{W}'$ in the proof above is biholomorphic.

3.5 Plücker Embedding

The Grassmann manifold $Gr(k, V)$ can be embedded into the complex projective space $\mathbb{P}(\Lambda^{n-k} V^*)$ as follows. Every k-dimensional subspace E of V is the zero locus of a set of $n - k$ linearly independent 1-forms $\xi_1, \ldots, \xi_{n-k}$ on V. Denote the projective class $[\xi_1 \wedge \ldots \wedge \xi_{n-k}]$ by $\jmath_{n-k}(E)$. This class does not depend on the choice of $\xi_1, \ldots, \xi_{n-k}$. For, let $\eta_1, \ldots \eta_{n-k}$ be linearly independent 1-forms such that $E = \{x \in V;\ \eta_1(x) = \cdots = \eta_{n-k}(x) = 0\}$. The two sets $\{\xi_i\}$ and $\{\eta_i\}$ give rise to two bases of $(V/E)^*$ in an obvious way, so $\eta_s = \sum_{r=1}^{n-k} a_{sr}\xi_r$, $s = 1, \ldots, n-k$, for some constants $a_{sr} \in \mathbb{C}$ with $\det[a_{sr}] \neq 0$. We have $\eta_1 \wedge \ldots \wedge \eta_{n-k} = \det[a_{sr}]\xi_1 \wedge \ldots \wedge \xi_{n-k}$, hence $[\eta_1 \wedge \ldots \wedge \eta_{n-k}] = [\xi_1 \wedge \ldots \wedge \xi_{n-k}]$. In this way, we get a map

$$\jmath_{n-k} : Gr(k, V) \to \mathbb{P}(\Lambda^{n-k} V^*),$$

called the **Plücker embedding**. We shall show that the map $\jmath_{n-k}$ is really a holomorphic embedding. We prove first that this map is holomorphic. For, let $\Omega_{A,W}$ be any standard coordinate neighbourhood of the Grassmann manifold. For $E \in \Omega_{A,W}$, let $u_E : A \to W$ be the linear map whose graph is E. Take bases $\{a_1, \ldots, a_k\}$ of A and $\{a_{k+1}, \ldots, a_n\}$ of W. The standard coordinates $\{z_{ir}\}$ of E are determined by the identities $u_E(a_i) = \sum_{r=k+1}^{n} z_{ir} a_r$, $i = 1, \ldots, k$. The vectors $b_i = a_i + \sum_{r=k+1}^{n} z_{ir} a_r$, $1 \le i \le k$, and $b_s = a_s$, $k+1 \le s \le n$, constitute a basis of V, $b_1, \ldots, b_k$ being a basis of E and $b_{k+1}, \ldots, b_n$ a basis of W. Let $\alpha_1, \ldots, \alpha_k$ be the dual basis of $a_1, \ldots, a_k$. Then

$$\beta_i = \alpha_i,\ 1 \le i \le k, \quad \beta_s = -\sum_{j=1}^{k} z_{js}\alpha_j + \alpha_s,\ k+1 \le s \le n \tag{3.12}$$

is the dual basis of $b_1, \ldots, b_n$. Clearly, $\beta_s(b_i) = 0$ for every $i = 1, \ldots, k$, so $\beta_s|E = 0$. Therefore if $p : (\Lambda^{n-k}V^*) \setminus \{0\} \to \mathbb{P}(\Lambda^{n-k}V^*)$ is the natural projection, the map $\jmath_{n-k}$ is given in the local coordinates $\{z_{ir}\}$ by $p(\beta_{k+1} \wedge \ldots \wedge \beta_n)$ where, according to (3.12), $\beta_{k+1} \wedge \ldots \wedge \beta_n$ holomorphically depends on $\{z_{ir}\}$. Therefore the local representation of $\jmath_{n-k}$ is a holomorphic map. This shows that the map $\jmath_{n-k}$ is holomorphic.

For every vector $v \in V$, denote by $\iota_v\varphi$ the interior product of v with a skew-symmetric form φ on the space V defined by $(\iota_v\varphi)(v_1, \ldots, v_{p-1}) = \varphi(v, v_1, \ldots, v_{p-1})$. Then if $\xi, \ldots, \xi_{n-k}$ are linear forms, $\iota_v(\xi_1 \wedge \ldots \wedge \xi_{n-k}) = \sum_{i=1}^{n-k} (-1)^{i-1}\xi_i(v)\xi_1 \wedge \ldots \wedge \widehat{\xi_i} \wedge \ldots \wedge \xi_{n-k}$. Therefore if $\jmath_{n-k}(E) = [(\xi_1 \wedge \ldots \wedge \xi_{n-k})]$, i.e., if ξ_i are linearly independent 1-forms whose zero locus is E, then $E = \{v \in V : \iota_v(\xi_1 \wedge \ldots \wedge \xi_{n-k}) = 0\}$. It follows that $\jmath_{n-k}$ is an injective map.

Next, we show that $\jmath_{n-k}$ is an immersion. For $g \in GL(V)$, denote the standard action $E \to g(E)$ of g on $Gr(k, V)$ again by g; the map $g : Gr(k, V) \to Gr(k, V)$ is biholomorphic. Let $g^* : \Lambda^{n-k}V^* \to \Lambda^{n-k}V^*$ be the isomorphism of the space of $(n-k)$- forms on V induced by g, so $(g^*\xi)(v_1, \ldots, v_{n-k}) = \xi(g(v_1), \ldots, g(v_{n-k}))$. Denote by $[g^*]$ the biholomorphism of $\mathbb{P}(\Lambda^{n-k}V^*)$ defined by $[g^*]([\xi]) = [g^*\xi]$. Then

$$\jmath_{n-k} \circ g = [g^*] \circ \jmath_{n-k}. \tag{3.13}$$

For every $E, F \in Gr(k, V)$, there is $g \in GL(V)$ such that $g(E) = F$, and $(d\jmath_{n-k})_F = (d[g^*])_{\jmath_{n-k}(E)} \circ (d\jmath_{n-k})_E \circ (dg^{-1})_F$ by (3.13). Since $d[g^*]$ and dg^{-1} are isomorphism at every point, we see that $rank\,(d\jmath_{n-k})_F = rank\,(d\jmath_{n-k})_E$. In other words, the rank of the differential $d\jmath_{n-k}$ is constant, say r. Then, by the constant rank theorem for holomorphic maps (see, e.g., Narasimhan [20], Chap. 2, § 2.4, Theorem 2.4.9), for every $E \in Gr(k, V)$, there are local coordinates of $Gr(k, V)$ around the point E and local coordinates of $\mathbb{P}(\Lambda^{n-k}V^*)$ around the point $\jmath_{n-k}(E)$ such that the representation of $\jmath_{n-k}$ in these coordinates is given by $(\zeta_1, \ldots, \zeta_{k(n-k)}) \to (\zeta_1, \ldots, \zeta_r, 0, \ldots, 0)$. This map is injective, since $\jmath_{n-k}$ is injective. Hence $r = k(n-k)$. Therefore $d\jmath_{n-k}$ is of rank equal to the dimension of $G(k, V)$, i.e., $\jmath_{n-k}$ is an immersion.

Consider the subset $M = \jmath_{n-k}((Gr(k, V))$ of $\mathbb{P}(\Lambda^{n-k}V^*)$ with the induced topology. The continuous map $\jmath_{n-k} : Gr(k, V) \to \mathbb{P}(\Lambda^{n-k}V^*)$ induces a continuous map

$\widetilde{\jmath}_{n-k} : Gr(k, V) \to M$. This map is bijective and its inverse is continuous. For, if $\mathcal{C}$ is a closed subset of $Gr(k, V)$, it is compact since the Grassmannian is a compact space. Then $\widetilde{\jmath}_{n-k}(\mathcal{C})$ is a compact subset of M, the latter space being Hausdorff since the projective space is so. Hence $\widetilde{\jmath}_{n-k}(\mathcal{C})$ is a closed subset of M. This proves the continuity of $\widetilde{\jmath}_{n-k}^{-1}$. In this way, we see that the map $\widetilde{\jmath}_{n-k} : Gr(k, V) \to M$ is a homeomorphism.

The arguments above prove the following.

Proposition 3.5.1 *The map $\jmath_{n-k} : Gr(k, V) \to \mathbb{P}(\Lambda^{n-k}V^*)$ is a holomorphic embedding.*

Problem 3.5.1 Find the coordinate representations of $\jmath_{n-k}$ and $d\jmath_{n-k}$ with respect to the standard local coordinates of $Gr(k, V)$ and $\mathbb{P}(\Lambda^{n-k}V^*)$.

Proposition 3.5.2 *Let Q be the tautological-quotient bundle over $Gr(k, V)$, and let $\mathcal{O}(1)$ be the dual bundle of the tautological bundle over $\mathbb{P}(\Lambda^{n-k}V^*)$. Then there is a fibre-preserving holomorphic map $\mathcal{J}_{n-k} : \Lambda^{n-k}Q \to \mathcal{O}(1)$ that is an isomorphism on each fibre and such that the diagramme*

$$\begin{array}{ccc} \Lambda^{n-k}Q & \xrightarrow{\mathcal{J}_{n-k}} & \mathcal{O}(1) \\ \downarrow & & \downarrow \\ Gr(k, V) & \xrightarrow[\jmath_{n-k}]{} & \mathbb{P}(\Lambda^{n-k}V^*) \end{array}$$

is commutative.

Proof Let $E \in Gr(k, V)$, and denote the equivalence class of a vector $v \in V$ in $Q_E = V/E$ by $\widetilde{v}$. The fibre $\mathcal{O}(1)_{\jmath_{n-k}(E)}$ is the dual space of the space $\mathcal{O}(-1)_{\jmath_{n-k}(E)}$. Take linearly independent 1-forms $\xi_1, \dots, \xi_{n-k}$ on V whose zero locus is E so that $\mathcal{O}(-1)_{\jmath_{n-k}(E)} = \mathbb{C}.\xi_1 \wedge \dots \wedge \xi_{n-k}$. For $v_1, \dots, v_{n-k} \in V$, define $\widehat{\mathcal{J}}^E_{n-k}(\widetilde{v}_1, \dots, \widetilde{v}_{n-k}) \in \mathcal{O}(-1)_{\mathcal{J}_{n-k}(E)}$ setting $\widehat{\mathcal{J}}^E_{n-k}(\widetilde{v}_1, \dots, \widetilde{v}_{n-k})(\xi_1 \wedge \dots \wedge \xi_{n-k}) = \det[\xi_l(v_m)]$. Clearly, $\widehat{\mathcal{J}}^E_{n-k}(\widetilde{v}_1, \dots, \widetilde{v}_{n-k})$ depends only on the equivalence classes $\widetilde{v}_i$, not on the choice of their representatives v_i. Also, it is skew-symmetric with respect to $\widetilde{v}_1, \dots, \widetilde{v}_{n-k}$, hence it gives rise to a linear map $\mathcal{J}^E_{n-k} : \Lambda^{n-r}Q_E \to \mathcal{O}(1)_{\jmath_{n-k}(E)}$ such that $\mathcal{J}^E_{n-k}(\widetilde{v}_1, \wedge \dots, \wedge \widetilde{v}_{n-k})(\xi_1 \wedge \dots \wedge \xi_{n-k}) = \det[\xi_l(v_m)]$. In order to find the kernel of the linear map $\mathcal{J}^E_{n-k}$, complete $\xi_1, \dots, \xi_{n-k}$ to a basis $\xi_1, \dots, \xi_n$ of the space V^*. Let $a_1, \dots, a_n$ be its dual vectors of V. The 1-forms $\xi_1, \dots, \xi_{n-k}$ have zero value for all vectors $a_{n-k+1}, \dots, a_n$, hence these vectors form a basis of E. Then $\widetilde{a}_1, \dots, \widetilde{a}_{n-k}$ is a basis of $Q_E = V/E$ thus, every vector $x \in \Lambda^{n-k}Q_E$ can be written as $x = \lambda\widetilde{a}_1 \wedge \dots \wedge \widetilde{a}_{n-k}$, $\lambda \in \mathbb{C}$. Hence if $\mathcal{J}^E_{n-k}(x) = 0$, then $0 = \lambda \det[\xi_l(a_m)]^{n-k}_{l,m=1} = \lambda$. Therefore $Ker\, \mathcal{J}^E_{n-k} = 0$. Since $dim\, \Lambda^{n-k}Q_E = dim\, \mathcal{O}(1)_{\jmath_{n-k}(E)} = 1$, the linear map $\mathcal{J}^E_{n-k} : \Lambda^{n-k}Q_E \to \mathcal{O}(1)_{\jmath(E)}$ is an isomorphism. The maps $\mathcal{J}^E_{n-k}$ give rise to a fibre-preserving map $\mathcal{J}_{n-k} : \Lambda^{n-k}Q \to \mathcal{O}(1)$.

So, what remains is to prove that the map $\mathcal{J}_{n-k} : \Lambda^{n-k}Q \to \mathcal{O}(1)$ is holomorphic. To this end, we shall represent this map in terms of frames. Let $\Omega_{A,W}$ be a standard coordinate neighbourhood of $Gr(k, V)$. Take a basis $a_1, \ldots, a_n$ of V such that $a_1, \ldots, a_k$ is a basis of A and $a_{k+1}, \ldots, a_n$ a basis of W. Let $\alpha_1, \ldots, \alpha_n$ be the dual basis of $a_1, \ldots, a_n$. Then $\jmath_{n-k}(A) = [\alpha_{k+1} \wedge \ldots \wedge \alpha_n]$. For $E \in \Omega_{A,W}$ and $l = 1, \ldots, n-k$, set $\sigma_l(E) =$ the equivalence class of a_{l+k} in $V/E = Q_E$. Then $\sigma = \sigma_1 \wedge \ldots \wedge \sigma_{n-k}$ is a holomorphic frame of $\Lambda^{n-k}Q$ on $\Omega_{A,W}$. Every element ξ of $\Lambda^{n-k}V^*$ can be uniquely written as $\xi = \sum_{1 \le i_1 < \ldots < i_{n-k}} \xi_{i_1,\ldots,i_{n-k}} \alpha_{i_1} \wedge \ldots \wedge \alpha_{i_{n-k}}$. Under this notation, $\jmath_{n-k}(A) = [\alpha_{k+1} \wedge \ldots \wedge \alpha_n]$ lies in the coordinate neighbourhood $U = \{[\xi] \in \mathbb{P}(\Lambda^{n-k}V^*) : \xi_{k+1,\ldots,n} \neq 0\}$. The frame $s([\xi]) = ([\xi], \dfrac{1}{\xi_{k+1,\ldots,n}}\xi)$ of the bundle $\mathcal{O}(-1)$ is defined on this neighbourhood. Let s^* be the dual frame of s. Then the holomorphicity of $\mathcal{J}_{n-k}$ would follow from the identity $\mathcal{J}_{n-k} \circ \sigma = s^* \circ \jmath_{n-k}$. To prove this identity, denote the standard coordinates of the points E of $\Omega_{A,W}$ by $\{z_{ir}\}$; they are determined via the identity $u_E(a_i) = \sum_{r=k+1}^{n} z_{ir} a_r$, $i = 1, \ldots, k$, where $u_E : A \to W$ is the linear map with graph E. Then, as noted above, the forms $\beta_s = -\sum_{j=1}^{k} z_{js}\alpha_j + \alpha_s$, $k+1 \le s \le n$, are linearly independent and their zero locus is E. Hence $\jmath_{n-k}(E) = [\beta_{k+1} \wedge \ldots \wedge \beta_n]$. Clearly, the coefficient of $\beta_{k+1} \wedge \ldots \wedge \beta_n$ in front of $\alpha_{k+1} \wedge \ldots \wedge \alpha_n$ in its decomposition with respect to the basis $\alpha_{i_1} \wedge \ldots \wedge \alpha_{i_{n-k}}$ of $\Lambda^{n-k}V^*$ is 1. Hence $\jmath_{n-k}(E) \in U$ and $s(\jmath_{n-k}(E)) = \beta_{k+1} \wedge \ldots \wedge \beta_n$. Then $s^*((\jmath_{n-k}(E))$ is the linear form that takes value 1 at the point $\beta_{k+1} \wedge \ldots \wedge \beta_n$. On the other hand, $\mathcal{J}_{n-k}(\sigma(E))$ is the linear form whose value at $\beta_{k+1} \wedge \ldots \wedge \beta_n$ is equal to $\det[\beta_s(a_r)]_{s,r=k+1}^{n} = \det[\delta_{sr}] = 1$. This proves the identity $\mathcal{J}_{n-k} \circ \sigma = s^* \circ \jmath_{n-k}$.

The map $\mathcal{J}_{n-k} : \Lambda^{n-k}Q \to \mathcal{O}(1)$ is holomorphic and is a vector space isomorphism on every fibre. □

Corollary 3.5.1 $\Lambda^{n-k}Q \cong \jmath_{n-k}^{*}\mathcal{O}(1)$.

3.6 The Canonical Bundle of a Complex Grassmann Manifold

Recall that the canonical bundle of a n-dimensional complex manifold X is the determinant bundle $det(T^hX)^* = \Lambda^n(T^hX)^*$. In order to compute the canonical bundle of the Grassmann manifold $Gr(k, V)$, we need the following general statement for determinant bundles.

Proposition 3.6.1 *Let $E \to M$, $F \to M$ be vector bundles of rank r and s, respectively. Then there is a canonical isomorphism*

$$det(E \otimes F) \cong (\otimes^s det\ E) \otimes (\otimes^r det\ F).$$

Proof Let $p \in M$ and let $\{e_1, \ldots, e_r\}$, $\{f_1, \ldots, f_s\}$ be bases of the fibres E_p and F_p, respectively. Then $\{e_i \otimes f_k : i = 1, \ldots, r, k = 1, \ldots, s\}$ is a basis of $E_p \otimes F_p$. Set

$$a = (e_1 \otimes f_1) \wedge (e_1 \otimes f_2) \wedge \ldots \wedge (e_1 \otimes f_s) \wedge (e_2 \otimes f_1) \wedge (e_2 \otimes f_2) \wedge \ldots \wedge (e_2 \otimes f_s) \\ \wedge \ldots \wedge (e_r \otimes f_1) \wedge (e_r \otimes f_2) \wedge \ldots \wedge (e_r \otimes f_s). \tag{3.14}$$

Clearly, a is a basis of the 1-dimensional vector space $\Lambda^{rs}(E_p \otimes F_p)$. Furthermore, the vector space $(\otimes^s \Lambda^r E_p) \otimes (\otimes^r \Lambda^s F_p)$ is also of dimension 1 with a basis

$$b = (\otimes^s (e_1 \wedge \ldots \wedge e_r)) \otimes (\otimes^r (f_1 \wedge \ldots \wedge f_s)). \tag{3.15}$$

Let $\chi_p : \Lambda^{rs}(E_p \otimes F_p) \to (\otimes^s \Lambda^r E_p) \otimes (\otimes^r \Lambda^s F_p)$ be the isomorphism defined by $\chi_p(a) = b$. We claim that this map does not depend on the choices of the bases $[e_i\}$ and $\{f_k\}$ of E_p and F_p. To prove this, let $\{e'_i\}$ and $\{f'_k\}$ be other bases of these spaces. Let a' and b' be defined by means of $\{e'_i\}$ and $\{f'_k\}$ via (3.14) and (3.15). Set $e'_i = \sum_{j=1}^r \lambda_{ij} e_j$, $f'_k = \sum_{l=1}^s \mu_{kl} f_l$. Then, for every $i = 1, \ldots, r$,

$$(e'_i \otimes f'_1) \wedge (e'_i \otimes f'_2) \wedge \ldots \wedge (e'_i \otimes f'_s)$$

$$= \sum_{l_1, l_2, \ldots, l_s = 1}^{s} \mu_{1l_1} \mu_{2l_2} \ldots \mu_{sl_s} (e'_i \otimes f_{l_1}) \wedge (e'_i \otimes f_{l_2}) \wedge \ldots \wedge (e_i \otimes f_{l_s}).$$

In fact, the summation is over different indices $l_1, l_2, \ldots, l_s$ since if $l_\alpha = l_\beta$ for some α, β, then $(e'_i \otimes f_{l_1}) \wedge (e'_i \otimes f_{l_2}) \wedge \ldots \wedge (e_i \otimes f_{l_s}) = 0$. If the s-tuple $(l_1, l_2, \ldots, l_s)$ consists of different numbers, let σ be the permutation sending $(1, 2, \ldots, s)$ to $(l_1, l_2, \ldots, l_s)$. Then

$$(e'_i \otimes f'_1) \wedge (e'_i \otimes f'_2) \wedge \ldots \wedge (e'_i \otimes f'_s)$$

$$\begin{aligned} &= \sum_\sigma \mu_{1\sigma(1)} \mu_{2\sigma(2)} \ldots \mu_{s\sigma(s)} (e'_i \otimes f_{\sigma(1)}) \wedge (e'_i \otimes f_{\sigma(2)}) \wedge \ldots \wedge (e_i \otimes f_{\sigma(s)}) \\ &= \sum_\sigma sign(\sigma) \mu_{1\sigma(1)} \mu_{2\sigma(2)} \ldots \mu_{s\sigma(s)} (e'_i \otimes f_1) \wedge (e'_i \otimes f_2) \wedge \ldots \wedge (e_i \otimes f_s) \\ &= det[\mu_{kl}] (e'_i \otimes f_1) \wedge (e'_i \otimes f_2) \wedge \ldots \wedge (e'_i \otimes f_s). \end{aligned}$$

Therefore

$$a' = (det[\mu_{kl}])^r (e'_1 \otimes f_1) \wedge \ldots \wedge (e'_1 \otimes f_s) \wedge \ldots \wedge (e'_r \otimes f_1) \wedge \ldots \wedge (e'_r \otimes f_s).$$

Let τ be the permutation of the set $\{(1, 1), (1, 2), \ldots, (1, s), \ldots, (r, 1), (r, 2), \ldots (r, s)\}$ reordering this set to $\{(1, 1), (2, 1), \ldots, (r, 1), \ldots, (1, s), (2, s), \ldots, (t, s)\}$. Then

$$\begin{aligned}
a' &= (det[\mu_{kl}])^r sign(\tau)(e_1' \otimes f_1) \wedge (e_2' \otimes f_1) \wedge \ldots \wedge (e_r' \otimes f_1) \\
&\quad \wedge \ldots \wedge (e_1' \otimes f_s) \wedge (e_2' \otimes f_s) \wedge \ldots \wedge (e_r' \otimes f_s) \\
&= (det[\mu_{kl}])^r sign(\tau)(det[\lambda_{ij}])^s (e_1 \otimes f_1) \wedge (e_2 \otimes f_1) \wedge \ldots \wedge (e_r \otimes f_1) \\
&\quad \wedge \ldots \wedge (e_1 \otimes f_s) \wedge (e_2 \otimes f_s) \wedge \ldots \wedge (e_r \otimes f_s) \\
&= (\det[\lambda_{ij}])^s (det[\mu_{kl}])^r (e_1 \otimes f_1) \wedge (e_1 \otimes f_2) \wedge \ldots \wedge (e_1 \otimes f_s) \\
&\quad \wedge \ldots \wedge (e_r \otimes f_1) \wedge (e_r \otimes f_2) \wedge \ldots \wedge (e_r \otimes f_s) \\
&= (\det[\lambda_{ij}])^s (det[\mu_{kl}])^r a.
\end{aligned}$$

Moreover, $e_1' \wedge \ldots \wedge e_r' = \det[\lambda_{ij}] e_1 \wedge \ldots \wedge e_r$ and $f_1' \wedge \ldots \wedge f_s' = det[\mu_{kl}] f_1 \wedge \ldots \wedge f_s$, hence

$$b' = (det[\lambda_{ij}])^s (\det[\mu_{kl}])^r b.$$

Therefore $\chi_p(a') = (\det[\lambda_{ij}])^s (det[\mu_{kl}])^r b = b'$. The maps χ_p yield a fibre preserving map $\chi : det(E \otimes F) \to (\otimes^s det\ E) \otimes (\otimes^r det\ F)$. By the very definition, χ maps frame to frame, hence it is a bundle isomorphism. □

Proposition 3.6.2 *The canonical bundle of the Grassmann manifold $Gr(k, V)$ is isomorphic to the bundle $\mathcal{O}(-n) \to \mathbb{P}(\Lambda^{n-k} V^*)$, where $\mathcal{O}(-1) \to \mathbb{P}(\Lambda^{n-k} V^*)$ is the tautological bundle.*

Proof Let S and Q be the tautological and tautological-quotient bundles over the Grassmann manifold. By Proposition 3.4.2, $T^h Gr(k, V) \cong S^* \otimes Q$, hence the cotangent bundle of the Grassmann manifold is $(T^h Gr(k, V))^* \cong S \otimes Q^*$. Then, by Proposition 3.6.1, $det\ Gr(k, V) \cong (\otimes^{n-k} det\ S) \otimes (\otimes^k det\ Q^*)$. By Chap. 1, Sect. 1.7, formula (1.14), the bundle $\underline{V}$ admits transition functions $g_{\alpha\beta}$ whose block-matrices are of the form

$$g_{\alpha\beta}(x) = \begin{bmatrix} g^S_{\alpha\beta}(x) & 0 \\ h_{\alpha\beta}(x) & g^Q_{\alpha\beta}(x) \end{bmatrix},$$

where $g^S_{\alpha\beta}$ and $g^Q_{\alpha\beta}$ are transition functions of the bundles S and Q, and $h_{\alpha\beta}(x)$ is a linear map $\mathbb{C}^{n-k} \to \mathbb{C}^k$. Then $det\ g_{\alpha\beta}$, $det\ g^S_{\alpha\beta}$, $det\ g^Q_{\alpha\beta}$ are transition functions of the bundles $det\ \underline{V}$, $det\ S$, $det\ Q$. The block-matrix form of $g_{\alpha\beta}$ implies $det\ g_{\alpha\beta} = det\ g^S_{\alpha\beta} . det\ g^Q_{\alpha\beta}$. This identity means that $det\ \underline{V} \cong det\ S \otimes det\ Q$. The determinant bundle of the trivial bundle $\underline{V}$ is isomorphic to the trivial line bundle L over the Grassmann manifold. Thus, $L \cong det\ S \otimes det\ Q$. By definition, $det\ Q^* = \Lambda^{n-k} Q^* \cong (\Lambda^{n-k} Q)^* = (det\ Q)^*$. Then

$$det\ Q^* \cong L \otimes det\ Q^* \cong det\ S \otimes det\ Q \otimes (det\ Q)^* \cong det\ S \otimes L \cong det\ S.$$

It follows that

$$det\, Gr(k, V) \cong (\otimes^{n-k} det\, S) \otimes (\otimes^{k} det\, S) \cong \otimes^{n} det\, S.$$

By Proposition 3.5.2, $det\, Q \cong \mathcal{O}(1)$. Hence $det\, S = (det\, Q)^* = \mathcal{O}(-1)$. It follows that

$$det\, Gr(k, V) \cong \mathcal{O}(-n).$$

□

3.7 Curvature of the Tautological and Tautological Quotient Bundles over a Complex Projective Space

Let V be a complex vector space with a Hermitian metric h, $dim V = n + 1$. This metric induces a metric on the trivial bundle $\underline{V} = \mathbb{P}(V) \times V$ and on its subbundle $\mathcal{O}(-1)$, the tautological bundle over $\mathbb{P}(V)$, in an obvious way; the induced metric will be denoted again by h.

Let $\xi \in \mathbb{P}(V)$. Take an orthonormal basis $v_0, v_1, \ldots, v_n$ of V such that $\xi = [v_0]$. Then ξ lies in the coordinate neighbourhood $U_0 = \{[v_0 + \sum_{k=1}^{n} z_k v_k] : z_1, \ldots, z_n \in \mathbb{C}\}$. In what follows, we work in the local coordinates $z_1, \ldots, z_n$ on U_0. As usual, the sections of the trivial bundle $\underline{V} = \mathbb{P}(V) \times V$ are identified with functions with values in V. Then $z = (z_1, \ldots, z_n) \to \varepsilon(z) = v_0 + z_1 v_1 + \cdots + z_n v_n$ is a nowhere vanishing holomorphic section of the line bundle $S = \mathcal{O}(-1)$. The connection form θ^S of the canonical connection D^S on S with respect to the frame ε is given by

$$\theta^S = \partial h(\varepsilon, \varepsilon).h(\varepsilon, \varepsilon)^{-1} = \partial \ln h(\varepsilon, \varepsilon) = \partial \ln (1 + |z|^2) = \frac{\sum_{k=1}^{n} \bar{z}_k dz_k}{1 + |z|^2}.$$

Therefore the curvature form with respect to ε is

$$\Theta^S = \bar{\partial}\theta^S = \sum_{k,j=1}^{n} \frac{\delta_{kj}(1 + |z|^2) - \bar{z}_k z_j}{(1 + |z^2)^2} d\bar{z}_j \wedge dz_k.$$

To compute the curvature of the quotient bundle $Q = \underline{V}/S$, denote the second fundamental form of S by A. The constant sections $v_k(z) = v_k$ of $\underline{V}$, $k = 1, \ldots, n$, together with ε constitute a holomorphic frame of $\underline{V}$. Then the sections $e_k = pr_{S^\perp}(v_k)$ form a holomorphic frame of Q. Clearly,

$$e_k = v_k - \frac{h(v_k, \varepsilon)}{h(\varepsilon, \varepsilon)} \varepsilon = v_k - \frac{\bar{z}_k}{1 + |z|^2} \varepsilon.$$

Set

$$e_0 = \frac{\varepsilon}{h(\varepsilon, \varepsilon)} = \frac{1}{1 + |z|^2} \varepsilon.$$

Then

$$D^{\underline{V}} e_0 = de_0 = d\frac{1}{1+|z|^2} \otimes \varepsilon + \frac{1}{1+|z|^2} d\varepsilon,$$

$$D^S e_0 = d\frac{1}{1+|z|^2} \otimes \varepsilon + \frac{1}{1+|z|^2} D^S \varepsilon,$$

where $D^S \varepsilon = \theta^S \otimes \varepsilon = \partial \ln(1+|z|^2) \otimes \varepsilon$. Hence

$$Ae_0 = D^{\underline{V}} e_0 - D^S e_0 = \frac{1}{1+|z|^2}\Big(\sum_{k=1}^{n} dz_k \otimes v_k - \frac{1}{1+|z|^2} \sum_{k=1}^{n} \overline{z}_k dz_k \otimes \varepsilon\Big). \quad (3.16)$$

Let $Ae_0 = \sum_{k=1}^{n} \lambda_k \otimes e_k$. Then

$$h(Ae_0, e_j) = \sum_{k=1}^{n} \lambda_k h(e_k, e_j) = \sum_{k=1}^{n} \lambda_k \Big(\delta_{kj} - \frac{\overline{z}_k z_j}{1+|z|^2}\Big).$$

On the other hand, by (3.16),

$$h(Ae_0, e_j) = \frac{1}{1+|z|^2}\Big(dz_j - \frac{z_j}{1+|z|^2} \sum_{k=1}^{n} \overline{z}_k dz_k\Big).$$

It follows that

$$\lambda_j = \frac{dz_j}{1+|z|^2}.$$

In accordance with the notations in Chap. 2, Sect. 2.8, set $Ae_0 = \sum_{k=1}^{n} a_{0k} \otimes e_k$. Then $a_{0k} = \dfrac{dz_k}{1+|z|^2}$, hence

$$\Theta^{Q}_{\nu\mu} = \Theta^{\underline{V}}_{\nu\mu} + a_{0\mu} \wedge \overline{a}_{0\nu} = \frac{dz_\mu \wedge d\overline{z}_\nu}{(1+|z|^2)^2}, \quad 1 \le \nu, \mu \le n.$$

Finally, note that, since $A \neq 0$, the bundle $\mathcal{O}(-1)^{\perp}$ which is isomorphic to Q as a smooth bundle is not a holomorphic subbundle of $\underline{V}$ (see the comments after Proposition 2.8.1 in Chap. 2).

Chern, Pontrjagin, and Euler Classes of Vector Bundles

4

4.1 De Rham Cohomology, A Short Overview

Let M be a smooth manifold and let $\Omega^p(M,\mathbb{K})$ be the space of differential p-forms on M with values in $\mathbb{K}$; if $p > \dim M$, then $\Omega^p(M,\mathbb{K}) = 0$. The exterior derivative d takes p-forms into $(p+1)$-forms. One of the main properties of the exterior derivative is that

$$d^2 = 0.$$

Recall that the forms φ for which $d\varphi = 0$ are called **closed** and those of the form $\psi = d\varphi$ - **exact**.

It is convenient to set $\Omega^{-1}(M,\mathbb{K}) = 0$. If $0 \le p \le n$, the quotient space

$$H^p_{dR}(M,\mathbb{K}) = (\ker d|\Omega^p(M,\mathbb{K}))/(d\Omega^{p-1}(M,\mathbb{K}))$$

is called the pth **de Rham cohomology group**; in fact, $H^p_{dR}(M,\mathbb{K})$ is a vector space over $\mathbb{K}$. Note that $H^0_{dR}(M,\mathbb{K})$ is the space of smooth $\mathbb{K}$-valued functions f on M such that $df = 0$, i.e., the space of locally constant $\mathbb{K}$-valued functions. If M is a connected manifold, $H^0_{dR}(M,\mathbb{K}) = \mathbb{K}$. Note also that $H^p_{dR}(M,\mathbb{C})$ is the complexification of $H^p_{dR}(M,\mathbb{R})$ since the space of complex-valued p-forms is the complexification of the space of real-valued p-forms.

Set

$$H^*_{dR}(M,\mathbb{K}) = \bigoplus_{p=0}^{\dim M} H^p_{dR}(M,\mathbb{K}).$$

Let ω and ω' be two closed forms and let $[\omega]$ and $[\omega']$ be their equivalence classes in the de Rham cohomology. Then the class $[\omega\wedge\omega']$ depends only on the classes $[\omega]$ and $[\omega']$, but not on their representatives ω and ω'. Indeed, for every two differential forms η and η'

$$\begin{aligned}(\omega+d\eta)\wedge(\omega'+d\eta') &= \omega\wedge\omega' + \omega\wedge d\eta' + d\eta\wedge\omega' + d\eta\wedge d\eta' \\ \text{''''} &= \omega\wedge\omega' + d((-1)^{\deg\omega}\omega\wedge\eta' + \eta\wedge\omega' + \eta\wedge d\eta'),\end{aligned}$$

J. Davidov, *Vector Bundles and Connections*, Compact Textbooks in Mathematics,
https://doi.org/10.1007/978-3-032-07403-4_4

since $d\omega = 0$ and $d\omega' = 0$. The cohomology class $[\omega \wedge \omega']$ is denoted by $[\omega].[\omega']$ or $[\omega] \cup [\omega']$ ("cup product"). It is easy to check that if ω'' is a closed differential form

$$([\omega] + [\omega']).[\omega''] = [\omega].[\omega''] + [\omega'].[\omega''].$$

Hence, $H^*_{dR}(M, \mathbb{K})$ admits a natural structure of a ring.

Example 4.1.1 By the classical Poincaré lemma

$$H^p_{dR}(\mathbb{R}^n, \mathbb{K}) = 0 \quad \text{for} \quad p \geq 1.$$

(see, e.g., Bott and Tu [2], Chap. I, § 4). Moreover, $H^0_{dR}(\mathbb{R}^n, \mathbb{K}) = \mathbb{K}$.

Example 4.1.2 The de Rham cohomology of the unit sphere is

$$H^p_{dR}(S^n, \mathbb{K}) = \begin{cases} \mathbb{K} \text{ if } & p = 0,\ n \\ 0 \text{ otherwise} \end{cases}$$

(Bott and Tu [2], Chap. I, § 4, Exercise 4.3).

Example 4.1.3 The de Rham cohomology groups of the real projective space are:

$$H^p_{dR}(\mathbb{RP}^n, \mathbb{K}) = \begin{cases} \mathbb{K} \text{ if } & p = 0 \\ 0 \text{ if } & 0 < p < n \\ \mathbb{K} \text{ if } & p = n \text{ and } n \text{ is odd} \\ 0 \text{ if } & p = n \text{ and } n \text{ is even} \end{cases}$$

(Bott and Tu [2], Chap. I, § 6, Exercise 6.46 (d)).

Example 4.1.4 For the complex projective space:

$$H^p_{dR}(\mathbb{CP}^n, \mathbb{K}) = \begin{cases} \mathbb{K} \text{ if } & p = 0, 2, 4, \ldots, 2n \\ 0 \text{ otherwise} \end{cases}$$

(see Bott and Tu [2], Chap. III, § 14, Example 14.22 and Exercise 14.22.1).

Example 4.1.5 Let M and N be smooth manifolds. By the Künneth formula

$$H^p_{dR}(M \times N, \mathbb{K}) = \bigoplus_{k+l=p} H^k_{dR}(M, \mathbb{K}) \otimes H^l_{dR}(N, \mathbb{K})$$

(Bott and Tu [2], Chap. I, § 5).

This formula can be used for computing the cohomology of $\mathbb{R}^n \setminus \{0\}$ taking into account the well-known fact that the manifold $\mathbb{R}^n \setminus \{0\}$ is diffeomorphic to $S^{n-1} \times \mathbb{R}$

(via the map $x \to (\frac{x}{|x|}, \ln|x|)$). Then, by Examples 4.1.1 and 4.1.2,

$$H^p_{dR}(\mathbb{R}^n \setminus \{0\}, \mathbb{K}) = \begin{cases} \mathbb{K}\otimes\mathbb{K} \cong \mathbb{K} & \text{if } \; p = 0, n-1 \\ 0 \text{ otherwise} \end{cases}$$

Every n-dimensional torus T^n is diffeomorphic to the product $S^1 \times ... \times S^1$, n factors. Using the Künneth formula and Example 4.1.2, it is easy to see by induction on n that

$$H^p_{dR}(T^n, \mathbb{K}) = \mathbb{K}^{\binom{n}{p}}$$

Let $f : M \to N$ be a smooth map of smooth manifolds. If $\varphi \in \Omega^p(N, \mathbb{K})$

$$d(f^*\varphi) = f^*(d\varphi).$$

Therefore the map $f^* : \Omega^p(N, \mathbb{K}) \to \Omega^p(M, \mathbb{K})$ induces a linear map

$$f^* : H^p_{dR}(N, \mathbb{K}) \to H^p_{dR}(M, \mathbb{K}), \quad [\omega] \to [f^*\omega].$$

Since $f^*(\omega \wedge \omega') = (f^*\omega) \wedge (f^*\omega')$, the map

$$f^* : H^*_{dR}(N, \mathbb{K}) \to H^*_{dR}(M, \mathbb{K})$$

is a morphism of rings. Clearly, if f is a diffeomorphism, f^* is an isomorphism.

Note that, by the de Rham theorem, de Rham cohomologies of a paracompact manifold are isomorphic to the singular cohomologies. The latter are topological invariants, homeomorphic manifolds have isomorphic singular cohomologies. Thus, although the de Rham cohomologies are defined by means of the smooth structures of manifolds, if two paracompact manifolds are homeomorphic, they have isomorphic de Rham cohomologies.

Recall that two smooth maps $f, g : M \to N$ are called homotopic if there is a smooth map $F : M \times [0, 1] \to N$ such that $F(x, 0) = f(x)$ and $F(x, 1) = g(x)$ for every $x \in M$. Two manifolds M and N are said to have the same homotopy type if there are smooth maps $f : M \to N$ and $g : N \to M$ such that the map $g \circ f$ is homotopic to Id_M and $f \circ g$ is homotopic to Id_N. It follows from the Poicaré lemma that homotopic maps induce the same maps of de Rham cohomologies (see, e.g., Bott and Tu [2], Chap. I, § 4, Corollary 4.1.2). It is then clear that manifolds of the same homotopy type have the same (i.e., isomorphic) de Rham cohomology groups.

An open cover $\{U_\alpha\}$ of a manifold M is called good if all non-empty finite intersections $U_{\alpha_0} \cap U_{\alpha_1} \cap \ldots \cap U_{\alpha_k}$ are diffeomorphic to the space $\mathbb{R}^n$. Every paracompact manifold admits a good cover. If the manifold is compact, it possesses a finite good cover (see Bott and Tu [2], Chap. I, § 5, Theorem 5.1).

If a manifold M admits a finite good cover, the spaces $H^p_{dR}(M, \mathbb{K})$ are finite-dimensional (Bott and Tu [2], Chap. I, § 5, Proposition 5.3.1). In particular, for every compact manifold

$$\dim H^p_{dR}(M, \mathbb{K}) < \infty.$$

Example 4.1.6 As we have noted, dim $H^2_{dR}(\mathbb{CP}^n, \mathbb{C}) = 1$. A basis element of the space $H^2_{dR}(\mathbb{CP}^n, \mathbb{C})$ is the cohomology class of the Kähler form ω of the Fubini-Study metric on $\mathbb{CP}^n$. For the definition and basic properties of the Fubini-Study metric, we refer to Moroianu [18], Part 2, Sects. 7.2 and 7.3 or Kobayashi and Nomizu [14], Vol. II, Chap. IX, § 6, Example 6.3. We recall here only the definition of the Kähler form ω. Let

$$U_l = \{[z_0, \ldots, z_n] \in \mathbb{CP}^n : z_l \neq 0\}, \quad l = 0, \ldots, n.$$

The functions $\zeta^l_j = \dfrac{z_j}{z_l}$, $\quad j = 0, \ldots, \widehat{l}, \ldots, n$, constitute a local coordinate system on U_l. Consider the locally defined $(1, 1)$-forms

$$\omega^l = \frac{i}{2}\partial\overline{\partial}\ln(|\zeta^l_0|^2 + \cdots + |\zeta^l_n|^2); \text{ here } \zeta^l_l = 1.$$

Note that the form ω^l takes real values for real tangent vectors, i.e., $\overline{\omega^l} = \omega^l$. Indeed, $\overline{\omega^l} = -\frac{i}{2}\overline{\partial}\partial\ln(|\zeta^l_0|^2 + \cdots + |\zeta^l_n|^2) = \omega^l$, since $\overline{\partial}\partial = -\partial\overline{\partial}$. Clearly $\zeta^k_j = \zeta^l_j \dfrac{1}{\zeta^l_k}$ on U_k, hence on $U_k \cap U_l$

$$\omega^k = \frac{i}{2}\partial\overline{\partial}\ln(\sum_{j=0}^{n} |\zeta^k_j|^2) = \frac{i}{2}\partial\overline{\partial}\ln\frac{\sum\limits_{j=0}^{n} |\zeta^l_j|^2}{|\zeta^l_k|^2} = \omega^l,$$

since for every non-vanishing holomorphic function f, we have $\partial\overline{\partial}\ln|f|^2 = 0$. Hence, the local forms ω^l glue together to determine a globally defined form ω. Since $d = \partial + \overline{\partial}$, $\partial^2 = \overline{\partial}^2 = 0$ and $\overline{\partial}\partial = -\partial\overline{\partial}$, we have $d\omega = 0$. On U_l,

$$\omega^l = \frac{i}{2}\sum\nolimits_{\alpha,\beta\neq l}\frac{\delta_{\alpha\beta}\sum\limits_{j=0}^{n}|\zeta^l_j|^2 - \zeta^l_\beta\overline{\zeta}^l_\alpha}{(\sum_{j=0}^{n}|\zeta^l_j|^2)^2} d\zeta^l_\alpha \wedge d\overline{\zeta}^l_\beta, \text{ where } \zeta^l_l = 1.$$

Further on, we shall consider the form

$$\omega_n = \frac{1}{\pi}\omega,$$

instead of the form ω. The volume form of the Fubini-Study metric h_{FS} is $\dfrac{\omega^n}{n!}$ and the volume of $\mathbb{CP}^n$ is the irrational number $\dfrac{\pi^n}{n!}$. If we consider the form ω_n and its corresponding metric $\dfrac{1}{\pi}h_{FS}$, then the volume of $\mathbb{CP}^n$ is $\dfrac{1}{n!}$, a rational number.

4.2 Invariant Polynomials

Let V be a finite-dimensional vector space over $\mathbb{K}$, $\dim V = n$, and let $S^p(V)$ be the space of symmetric $\mathbb{K}$-multilinear maps $f : V \times \ldots \times V \to \mathbb{K}$, p factors. Recall that the direct sum $S(V) = \bigoplus_{p=0}^{\infty} S^p(V)$ has the structure of an algebra over the field $\mathbb{K}$, the multiplication of $f \in S^p(V)$ and $g \in S^q(V)$ being defined by

$$(f \odot g)(v_1, \ldots, v_{p+q}) = \frac{1}{p!q!} \sum_{\sigma} f(v_{\sigma(1)}, \ldots, v_{\sigma(p)}) g(v_{\sigma(p+1)}, \ldots, v_{\sigma(p+q)}),$$

where the sum is taken over all permutations σ of the numbers $\{1, 2, \ldots, p+q\}$.

A map $\varphi : V \to \mathbb{K}$ is called a homogeneous polynomial function of degree k if it is a homogeneous polynomial of degree k of n linear functions $\lambda_i : V \to \mathbb{K}$ with coefficients in $\mathbb{K}$, i.e., if it can be written as

$$\varphi(v) = \sum_{i_1,\ldots,i_k=1}^{n} a_{i_1,\ldots,i_k} \lambda_{i_1}(v) \ldots \lambda_{i_k}(v),$$

where the coefficients $a_{i_1,\ldots,i_k} \in \mathbb{K}$ are symmetric with respect to the indices $(i_1, \ldots, i_k)$. If $\xi^1, \ldots, \xi^n$ is a basis of the dual space V^*, the map φ is a homogeneous polynomial function of degree k if and only if it can be expressed as a homogeneous polynomial of degree k of $\xi^1, \ldots, \xi^n$. A map $\varphi : V \to \mathbb{K}$ is called a polynomial function (shortly "a polynomial on V") if it is a sum of homogeneous polynomial functions.

Denote by $P^k(V)$ the space of homogeneous polynomial functions of degree k on V. Then $P(V) = \bigoplus_{k=0}^{\infty} P^k(V)$ is the algebra of polynomial functions on V.

For a symmetric form $f \in S^k(V)$, define a function $\widetilde{f} : V \to \mathbb{K}$ by

$$\widetilde{f}(v) = f(v, \ldots, v).$$

Let $e_1, \ldots, e_n$ be a basis of V and $\xi^1, \ldots, \xi^n$ its dual basis. Then

$$\widetilde{f}(v) = \sum_{i_1,\ldots,i_k=1}^{n} f(e_{i_1}, \ldots, e_{i_k}) \xi^{i_1}(v) \ldots \xi^{i_k}(v).$$

Hence, $\widetilde{f} \in P^k(V)$. The correspondence $f \to \widetilde{f}$ is an isomorphism of $S(V)$ onto $P(V)$. The inverse map is given as follows. Every $\varphi \in P^k(V)$ is of the form

$$\varphi = \sum_{i_1,\ldots,i_k=1}^{n} a_{i_1,\ldots,i_k} \xi^{i_1} \ldots \xi^{i_k},$$

where the coefficients $a_{i_1,\dots,i_k} \in \mathbb{K}$ are symmetric with respect to $(i_1, \dots, i_k)$. Set

$$f(v_1, \dots, v_k) = \sum_{i_1,\dots,i_k=1}^{n} a_{i_1,\dots,i_k} \xi^{i_1}(v_1) \dots \xi^{i_k}(v_k).$$

Then $\widetilde{f} = \varphi$.

Let $\omega_1, \dots, \omega_p$ be skew-symmetric forms on a smooth manifold M with values in V of degrees $d_1, \dots, d_p$. For $f \in S^p(V)$, define a $\mathbb{K}$-valued skew-symmetric form $f(\omega_1, \dots, \omega_p)$ on M of degree $d_1 + \dots + d_p$ setting for $X_1, \dots, X_{d_1+\dots+d_p} \in T_x M$

$$f(\omega_1, \dots, \omega_p)(X_1, \dots, X_{d_1+\dots+d_p})$$
$$= \frac{1}{d_1! \dots d_p!} \sum_{\sigma} \epsilon(\sigma) f(\omega_1(X_{\sigma(1)}, \dots, X_{\sigma(d_1)}), \dots, \omega_p(X_{\sigma(d_1+\dots+d_{p-1}+1)}, \dots, X_{\sigma(d_1+\dots+d_p)})), \tag{4.1}$$

where the sum is taken over all permutations σ of $\{1, \cdots, d_1 + \cdots + d_p\}$ and $\epsilon(\sigma)$ stands for the signature of σ. The form $f(\omega_1, \dots, \omega_p)$ is the anti-symmetrization of the form

$$(X_1, \dots, X_{d_1+\dots+d_p}) \to f(\omega_1(X_1, \dots, X_{d_1}), \dots, \omega_p(X_{d_1+\dots+d_{p-1}+1}, \dots, X_{d_1+\dots+d_p})).$$

Let $e_1, \dots, e_n$ be a basis of V and $\xi^1, \dots, \xi^n$ its dual basis of V^*. Let

$$f(v_1, \dots, v_p) = \sum_{i_1,\dots,i_p=1}^{n} a_{i_1,\dots,i_p} \xi^{i_1}(v_1) \dots \xi^{i_p}(v_p).$$

Write $\omega_l = \sum_{i=1}^{n} \omega_l^i e_i$, where ω_l^i are $\mathbb{K}$-valued differential forms on M of degree d_l. Under this notation, the form $f(\omega_1, \dots, \omega_p)$ is the anti-symmetrization of

$$(X_1, \dots, X_{d_1+\dots+d_p}) \to$$
$$\sum_{i_1,\dots,i_p=1}^{n} a_{i_1,\dots,i_p} \xi^{i_1}(\omega_1(X_1, \dots, X_{d_1})) \dots \xi^{i_p}(\omega_p(X_{d_1+\dots+d_{p-1}+1}, \dots, X_{d_1+\dots+d_p}))$$
$$= \sum_{i_1,\dots,i_p=1}^{n} a_{i_1,\dots,i_p} \omega_1^{i_1}(X_1, \dots, X_{d_1}) \dots \omega_p^{i_p}(X_{d_1+\dots+d_{p-1}+1}, \dots, X_{d_1+\dots+d_p})$$

It follows

$$f(\omega_1, \dots, \omega_p) = \sum_{i_1,\dots,i_p=1}^{n} a_{i_1,\dots,i_p} \omega_1^{i_1} \wedge \dots \wedge \omega_p^{i_p}. \tag{4.2}$$

If $\omega_1, \dots, \omega_p$ are sums of skew-symmetric forms of different degrees. i.e., if $\omega_1, \dots, \omega_p \in A^*(V) = \bigoplus_{d=0}^{\dim M} A^d(V)$, we can define $f(\omega_1, \dots, \omega_p)$ by means of formula (4.1) extended on $A^*(V)$ by multilinearity. It is then clear that formula (4.2) holds for all forms in $A^*(V)$.

Let G be a group of linear transformations of V. A form $f \in S^p(V)$ is said to be **invariant** with respect to G, if

$$f(gv_1, \dots, gv_p) = f(v_1, \dots, v_p)$$

for every $g \in G$ and $v_1, \dots, v_p \in V$. Similarly, a polynomial function $\varphi \in P^k(V)$ is called **invariant** if $\varphi(gv) = \varphi(v)$ for every $g \in G$ and $v \in V$. The correspondence $f \to \widetilde{f}$ defined above is an isomorphism of the algebra of invariant symmetric multilinear forms onto the algebra of invariant polynomial functions.

Now, let $V = \mathcal{M}_n$ be the space of $(n \times n)$-matrices with entries in $\mathbb{K}$. The group $\mathrm{GL}(n, \mathbb{K})$ of non-singular $(n \times n)$-matrices G acts on $\mathcal{M}_n$ by conjugation:

$$G.A = GAG^{-1}, \quad A \in \mathcal{M}_n, \quad G \in \mathrm{GL}(\mathbb{K}).$$

Further on, we consider invariant polynomial functions on $\mathcal{M}_n$.

Example 4.2.1 For $A \in \mathcal{M}_n$, set

$$\det(A + tI) = P^0(A)t^n + P^1(A)t^{n-1} + \dots + P^{n-1}(A)t + P^n(A),$$

where I is the unit $(n \times n)$-matrix and $t \in \mathbb{K}$ is a parameter. The coefficients $P^k(A)$, $k = 0, 1, \dots, n$, are invariant polynomial functions of A. In particular, $P^0(A) = 1$, $P^1(A) = Trace\, A$, $P^n(A) = \det A$.

It is clear that in the case when $\mathbb{K} = \mathbb{R}$ the polynomial functions $P^k(A)$ take real values.

In the case $\mathbb{K} = \mathbb{C}$, every matrix $A \in \mathcal{M}_n$ possesses eigenvalues, the roots of the characteristic equation $\det(A - tI) = 0$. Hence, $P^k(A)$ is the kth elementary symmetric function of the eigenvalues of A (Vieta's formulas). Let these eigenvalues be $\alpha_1, \dots, \alpha_n$. Then

$$\begin{aligned} P^2(A) &= \tfrac{1}{2} \sum_{i \neq j} \alpha_i \alpha_j = \tfrac{1}{2} \left(- \sum_{i=1}^{n} \alpha_i^2 + \sum_i \alpha_i \sum_j \alpha_j \right) \\ &= \tfrac{1}{2} \left(-Trace\, A^2 + (Trace\, A)^2 \right). \end{aligned}$$

We claim that for every k

$$P^k(A) = Trace\,(\Lambda^k A). \tag{4.3}$$

To prove this identity, it is enough to prove it in the case when A is a Jordan matrix. Indeed, let $A^{'} = GAG^{-1}$, $G \in \mathrm{GL}(n, \mathbb{C})$, be the Jordan normal form of A. Since $\Lambda^k A^{'} = \Lambda^k G \,.\, \Lambda^k A \,.\, \Lambda^k G^{-1} = \Lambda^k G \,.\, \Lambda^k A \,.\, (\Lambda^k G)^{-1}$, we have $Trace\,(\Lambda^k A^{'}) = Trace\,(\Lambda^k A)$. Moreover, $P^k(A^{'}) = P^k(A)$. Hence, if the identity $P^k(A^{'}) = Trace\,(\Lambda^k A^{'})$ holds, then (4.3) would hold. Thus, let A be a Jordan matrix with eigenvalues $\alpha_1, \dots, \alpha_n$. Let $e_1, \dots, e_n$ be the standard basis of $\mathbb{C}^n$. Then $e_{i_1} \wedge \dots \wedge e_{i_k}$, $1 \leq i_1 < \dots < i_k \leq n$, is a basis of $\Lambda^k \mathbb{C}^n$. Note also that $(\Lambda^k A)(e_{i_1} \wedge \dots \wedge e_{i_k}) = A(e_{i_1}) \wedge \dots \wedge A(e_{i_k})$. For every $i = 1, \dots, n$, we have either $A(e_i) = \alpha_i e_i + e_{i+1}$ or $A(e_i) = \alpha_i e_i$. Therefore the coefficient of $e_{j_1} \wedge \dots \wedge e_{j_k}$ in the representation of $(\Lambda^k A)(e_{j_1} \wedge \dots \wedge e_{j_k})$ with respect to the basis $e_{i_1} \wedge \ldots \wedge e_{i_k}$ is $\alpha_{j_1} \ldots \alpha_{j_k}$, $1 \leq j_1 < \ldots < j_k \leq n$. This shows that $Trace\,(\Lambda^k A) = \sum_{i_1 < \ldots < i_k} \alpha_{i_1} \ldots \alpha_{i_k}$, the kth elementary symmetric function. The latter equals $P^k(A)$ as we have noted.

Let g be the standard Hermitian metric of $\mathbb{C}^n$. Denote the induced metric on $\Lambda^k \mathbb{C}^n$ by g_k. Then if $A = [a_{ij}]$

$$
\begin{aligned}
Trace\,(\Lambda^k A) &= \sum_{1 \leq i_1 < \ldots < i_k \leq n} g_k(A(e_{i_1}) \wedge \ldots \wedge A(e_{i_k}), e_{i_1} \wedge \ldots \wedge e_{i_k}) \\
&= \sum_{1 \leq i_1 < \ldots < i_k \leq n} \det[g(A(e_{i_p}), e_{i_q})]_{p,q=1}^k = \sum_{1 \leq i_1 < \ldots < i_k \leq n} \det[a_{i_p, i_q}].
\end{aligned}
$$

For every multiindex $\mathcal{I} = \{i_1, \ldots, i_k\}$ with length k, $1 \leq i_1 < \ldots < i_k \leq n$, denote by $A_{\mathcal{I}}$ the minor $[a_{lm}]_{l,m \in \mathcal{I}}$ of the matrix $A = [a_{ij}]_{i,j}$. Then the identity above and (4.3) imply

$$
P^k(A) = \sum_{length(\mathcal{I})=k} \det(A_{\mathcal{I}}).
$$

Remark 4.2.1 Another proof of identity (4.3) can be given by means of the observation that if the matrix A is diagonalizable, then $Trace\,(\Lambda^k A)$ is the $k$the symmetric function of its eigenvalues. Hence, identity (4.3) holds for diagonalizable matrices. The set of diagonalizable matrices is a dense subset of $\mathcal{M}_n$ and (4.3) follows from the continuity of the function $A \to P^k(A)$.

Here is a proof of the claim that the set of diagonalizable matrices is a dense subset of $\mathcal{M}_n$ in the case $\mathbb{K} = \mathbb{C}$ (if $\mathbb{K} = \mathbb{R}$ this is not true). Since the field $\mathbb{C}$ is algebraically closed, every $n \times n$-matrix A with complex entries is conjugate to a triangular matrix B: $A = CBC^{-1}$ for some matrix C. For every sufficiently small $\varepsilon > 0$, we can change the diagonal entries of B so that the resulting triangular matrix B_ε to have different diagonal entries and to converge to B when $\varepsilon \to 0$ (give the details). It is clear that the eigenvalues of B_ε are different. The eigenvalues of the matrix $CB_\varepsilon C^{-1}$ coincide with that of B_ε. Thus, the matrix $CB_\varepsilon C^{-1}$ has different eigenvalues, hence it is diagonalizable. Clearly, $CB_\varepsilon C^{-1} \to CBC^{-1} = A$.

We note finally that every invariant polynomial function can be represented as a polynomial of the elementary symmetric polynomials P^k, $k = 0, \ldots, n$, see Milnor and Stasheff [17], Appendix C, Lemma 6).

4.3 Chern-Weil Homomorphism

Let $E \to M$ be a smooth $\mathbb{K}$-bundle and $\varphi \in A^d(Hom(E, E))$. Fix a point $x \in M$ and take a basis $e_1, \ldots, e_r$ of the fibre E_x. For tangent vectors $X_1, \ldots, X_p \in T_x M$, set

$$
\varphi(X_1, \ldots, X_p)e_i = \sum_{j=1}^{r} \varphi_{ij}(X_1, \ldots, X_p)e_j.
$$

Denote by $L_{ij} : E_x \to E_x$ the linear map determined by

$$L_{ij}(e_i) = e_j, \quad L_{ij}(e_k) = 0 \quad \text{for } k \neq i.$$

Then $\{L_{ij}\}$ is a basis of $Hom(E_x, E_x)$ and

$$\varphi(x) = \sum_{i,j=1}^{r} \varphi_{ij} \otimes L_{ij}.$$

Denote the matrix of the p-forms φ_{ij} by Φ. If $e'_1, \ldots, e'_r$ is another basis of E_x and $e'_i = \sum\limits_j g_{ij} e_j$, then

$$\Phi' = g \Phi g^{-1}, \tag{4.4}$$

where $g = [g_{ij}]$.

Now, let $F \in S^p(\mathcal{M}_r)$ be an invariant symmetric p-form. Given forms $\varphi^k \in A^{d_k}(Hom(E, E))$, $k = 1, \ldots, p$, we can define a $\mathbb{K}$-valued differential form on M of degree $(d_1 + \cdots + d_p)$ in the following way. Let $x \in M$ and let $e_1, \ldots, e_r$ be a basis of E_x. Denote by Φ^k the matrix $[\varphi^k_{ij}]_{i,j}$ of differential d_k-forms representing the form φ^k by means of the basis $e_1, \ldots, e_r$. Then set

$$F(\varphi^1, \ldots, \varphi^p)|_x = F(\Phi^1, \ldots, \Phi^p).$$

The right-hand side of this identity is understood in the sense of (4.1) (with $V = \mathcal{M}_r$, $\omega^k = \Phi^k$, $f = F$). It does not depend on the choice of the basis $e_1, \ldots, e_r$ since changing the basis changes the matrices Φ^i by the rule (4.4) and F is an invariant form.

For the invariant form F, there are constants $\lambda_{i_1,\ldots,i_p,j_1,\ldots,j_p} \in \mathbb{K}$ such that if $A^1, \ldots, A^p \in \mathcal{M}_r$ and $A^k = [a^k_{ij}]$, the form $F(A^1, \ldots, A^p)$ has the representation

$$F(A^1, \ldots, A^p) = \sum_{\substack{i_1,\ldots,i_p \\ j_1,\ldots,j_p}} \lambda_{i_1,\ldots,i_p,j_1,\ldots,j_p} a^1_{i_1 j_1} \cdots a^p_{i_p j_p},$$

where the coefficients $\lambda_{i_1,\ldots,i_p,j_1,\ldots,j_p}$ are symmetric in the pairs (i_1, j_i), ..., (i_p, j_p). Then, as we have noted in the preceding section,

$$F(\varphi^1, \ldots, \varphi^p) = \sum \lambda_{i_1,\ldots,i_p,j_1,\ldots,j_p} \varphi^1_{i_1 j_1} \wedge \ldots \wedge \varphi^p_{i_p j_p}. \tag{4.5}$$

This formula holds also for $\varphi^1, \ldots, \varphi^p \in \bigoplus\limits_{d=0}^{\dim M} A^d(Hom(E, E))$.

Now, given a connection D on the bundle E, we derive a useful formula expressing the differential of the form $F(\varphi^1, \ldots, \varphi^p)$ by the differentials of the forms $D\varphi^1, \ldots, D\varphi^p$, the differential operators being determined by D.

Let first $\omega_1, \ldots, \omega_p$ be differential forms on M of degrees $d_1, \ldots, d_p$. Then, setting $d_0 = 0$,

$$\begin{aligned} d(\omega_1 \wedge \ldots \wedge \omega_p) &= d\omega_1 \wedge \omega_2 \wedge \ldots \wedge \omega_p + (-1)^{d_1} \omega_1 \wedge d(\omega_2 \wedge \ldots \wedge \omega_p) \\ &= d\omega_1 \wedge \omega_2 \wedge \ldots \wedge \omega_p \\ &+ (-1)^{d_1} (\omega_1 \wedge d\omega_2 \wedge \omega_3 \wedge \ldots \wedge \omega_p \\ &\qquad + (-1)^{d_2} \omega_1 \wedge \omega_2 \wedge d(\omega_3 \wedge \ldots \wedge \omega_p)) \\ &= \ldots\ldots\ldots\ldots\ldots\ldots\ldots\ldots\ldots\ldots\ldots\ldots \\ &= \sum_{k=1}^{p} (-1)^{d_0 + \cdots + d_{k-1}} \omega_1 \wedge \ldots \wedge d\omega_k \wedge \ldots \wedge \omega_p, \end{aligned}$$

Take a frame $e_1, \ldots, e_r$ of E and represent $F(\varphi^1, \ldots, \varphi^p)$ in the form (4.5). Then

$$dF(\varphi^1, \ldots, \varphi^p) = \sum_{\substack{i_1, \ldots, i_p \\ j_1, \ldots, j_p}} \lambda_{i_1, \ldots, i_p, j_1, \ldots, j_p} \sum_{k=1}^{p} (-1)^{d_0 + \cdots + d_{k-1}} \varphi^1_{i_1 j_1} \wedge \ldots \wedge d\varphi^k_{i_k j_k} \wedge \ldots \wedge \varphi^p_{i_p j_p}. \tag{4.6}$$

Suppose we are given a connection D on the bundle E. Fix a point $x \in M$ and take a frame $e_1, \ldots, e_r$ in a neighbourhood of x such that $De_i|_x = 0$. Denote the connection on the bundle $Hom(E, E)$ induced by D again by D. The sections L_{ij} of $Hom(E, E)$ defined by $L_{ij}(e_k) = \delta_{ik} e_j,\ k = 1, \ldots, r$, constitute a frame of the bundle $Hom(E, E)$. For this frame

$$(DL_{ij})(e_k) = D(\delta_{ik} e_j) - L_{ij}(De_k) = \delta_{ik} De_j - L_{ij}(De_k).$$

Hence,

$$DL_{ij}|_x = 0.$$

since $De_l|_x = 0$ for every $l = 1, \ldots, r$.

Every form $\varphi \in A^m(\mathrm{Hom}(E, E))$ has the local representation

$$\varphi = \sum_{i,j} \varphi_{ij} \otimes L_{ij},$$

where φ_{ij} are $\mathbb{K}$-valued differential m-forms. Let $\theta_{ij,kl}$ be the forms of the connection D on $Hom(E, E)$ with respect to the frame L_{ij}: $DL_{ij} = \sum_{kl} \theta_{ij,kl} \otimes L_{kl}$. Then by Chap. 2, Sect. 2.6, identity (2.9),

$$D\varphi = \sum_{k,l} [d\varphi_{kl} + (-1)^m \sum_{i,j} \varphi_{ij} \wedge \theta_{ij,kl}] \otimes L_{kl}.$$

In this identity, $\theta_{ij,kl}|_x = 0$ since $DL_{ij}|_x = 0$. Hence, at the point x,

$$D\varphi|_x = \sum_{k,l} d\varphi_{kl} \otimes L_{kl}|_x. \tag{4.7}$$

Thus, at x, $D\varphi$ is represented by the matrix $[d\varphi_{ij}]|_x$. It follows from (4.6) that at the point x

$$dF(\varphi^1, \dots, \varphi^p) = \sum_{k=1}^{p} (-1)^{d_0+\cdots+d_{k-1}} F(\varphi^1, \dots, D\varphi^k, \dots, \varphi^p). \tag{4.8}$$

Proposition 4.3.1 *Let $R \in A^2(Hom(E, E))$ be the curvature tensor of a connection D on a bundle $E \to M$ of rank r. Let P be an invariant polynomial function on $\mathcal{M}_r$ of degree p. Then*

(1) The form $P(R)$ on M is closed.

(2) The cohomology class $[P(R)] \in H^{2p}_{dR}(M, \mathbb{K})$ does not depend on the choice of the connection D on E.

Proof (1) Let $x \in M$ and take a frame $e_1, \dots, e_r$ of E such that $De_i|_x = 0$.

Let $F \in S^p(\mathcal{M}_r)$ be the symmetric invariant form corresponding to the polynomial function P. Then, by (4.8),

$$dP(R) = dF(R, \dots, R) = \sum_{k=1}^{p} F(\underbrace{R, \dots, DR}_{k}, \dots, R).$$

If Θ is the curvature matrix of D with respect to the frame $e_1, \dots, e_r$, then $R = \sum_{i,j} \Theta_{ij} \otimes L_{ij}$ (Chap. 2, Sect. 2.7). Hence, by (4.7), the form DR at the point x is represented by the matrix $d\Theta$. Therefore, at x,

$$dP(R) = \sum_{k=1}^{p} F(\Theta, \dots, d\Theta, \dots, \Theta). \tag{4.9}$$

Let θ be the connection matrix of D with respect to the frame $e_1, \dots, e_r$. Then

$$\Theta = d\theta - \theta \wedge \theta,$$

hence $d\Theta = -d\theta \wedge \theta + \theta \wedge d\theta$. Since $\theta|_x = 0$, $d\Theta|_x = 0$ and it follows from (4.9) that

$$dP(R)|_x = 0.$$

(2) Let D and $\widetilde{D}$ be two connections on E. Consider the connections

$$D_t = (1-t)D + t\widetilde{D}, \quad 0 \le t \le 1,$$

joining D and $\widetilde{D}$. Write $D_t = D + t(\widetilde{D} - D)$ and set

$$B = \widetilde{D} - D.$$

If f is a smooth function on M and s is a section of E,

$$B(fs) = df \otimes s + f\widetilde{D}s - df \otimes s - fDs = fB(s).$$

Therefore $D - \widetilde{D}$ determines a tensor $B \in A^1(\mathrm{Hom}(E, E))$ (see Chap. 1, Example 1.6.6).

Let $R_t = R^{D_t} \in A^1(Hom(E, E))$ be the curvature tensor of the connection D_t. We are going to prove that

$$\frac{d}{dt}P(R_t) = p\,dF(B, R_t, \ldots, R_t), \quad 0 \le t \le 1.$$

Let $x \in M$. Fix t and take a frame $e_1, \ldots, e_r$ of E in a neighbourhood of x such that $D_t e_i|_x = 0, \quad i = 1, \ldots, r$. For $\tau \in [0, 1]$, denote by θ_τ and Θ_τ the connection and curvature matrices of D_τ with respect to the frame $e_1, \ldots, e_r$. Let θ be the connection matrix of D with respect to this frame, and β the matrix representing B in it. Since $D_\tau = D + \tau B$,

$$\theta_\tau = \theta + \tau\beta.$$

Then

$$\Theta_\tau = d\theta_\tau - \theta_\tau \wedge \theta_\tau = d(\theta + \tau\beta) - (\theta + \tau\beta) \wedge (\theta + \tau\beta).$$

This gives

$$\begin{aligned} \frac{d\Theta_\tau}{d\tau}\bigg|_{\tau=t} &= d\beta - \beta \wedge (\theta + t\beta) - (\theta + t\beta) \wedge \beta \\ &= d\beta - \beta \wedge \theta_t - \theta_t \wedge \beta. \end{aligned}$$

Since $\theta_t|_x = 0$, we get at the point x

$$\frac{d\Theta_\tau}{d\tau}\bigg|_{\tau=t} = d\beta.$$

Note also that, by (4.7), $D_t B$ at x is represented by the matrix $d\beta$. Since the form F is symmetric, applying (4.8), we obtain at x

$$\begin{aligned} \frac{d}{d\tau}P(R_\tau)\bigg|_{\tau=t} &= \frac{d}{d\tau}F(\Theta_\tau, \ldots, \Theta_\tau)\bigg|_{\tau=t} \\ &= pF(\frac{d\Theta_\tau}{d\tau}\bigg|_{\tau=t}, \Theta_t, \ldots, \Theta_t) \\ &= pF(d\beta, \Theta_t, \ldots, \Theta_t) \\ &= pF(D_t B, R_t, \ldots, R_t). \end{aligned}$$

On the other hand, we have $0 = D_t e_i|_x = \sum_{j=1}^{r} (\theta_t)_{ij} \otimes e_j|_x$. Thus, $\theta_t = [(\theta_t)_{ij}] = 0$ at x and the identity $d\Theta_t = -d\theta_t \wedge \theta_t + \theta_t \wedge d\theta_t$ gives $d\Theta_t|_x = 0$. By (4.7), the matrix $d\Theta_t|_x$ represents the form $D_t R_t|_x$ with respect to the frame $e_1, \ldots, e_r$. Then, by (4.8), at x,

$$\begin{aligned}dF(B, R_t, \dots, R_t) &= F(D_t B, R_t, \dots, R_t) + F(B, D_t R_t, \dots, R_t) + \cdots \\ &\quad + F(B, R_t, \dots, D_t R_t) \\ &= F(D_t B, R_t, \dots, R_t) + F(\beta, d\Theta_t, \dots, \Theta_t) + \cdots \\ &\quad + F(\beta, \Theta_t, \dots, d\Theta_t) \\ &= F(D_t B, R_t, \dots, R_t).\end{aligned}$$

It follows that

$$\frac{d}{dt} P(R_t) = p\, dF(B, R_t, \dots, R_t).$$

Integrating this identity on the interval [0, 1], we obtain

$$P(R_1) - P(R_0) = \int_0^1 p\, dF(B, R_t, \dots, R_t) dt = d \int_0^1 p\, F(B, R_t, \dots, R_t) dt.$$

Clearly, in the latter identity

$$\omega = \int_0^1 pF(B, R_t, \dots, R_t) dt$$

is a $(2p - 1)$-form on M. Finally, note that

$$P(R_1) = P(R^{\widetilde{D}}) \text{ and } P(R_0) = P(R^D).$$

□

Remark 4.3.1 Instead of invariant polynomial functions, one can use invariant formal power series of the form

$$P = P_0 + P_1 + P_2 + \cdots$$

where P_k is an invariant polynomial function of degree k on M. Then $P(R)$ is a well-defined form since $P_k(R) = 0$ for $2k > \dim M$. An important example of an invariant power series is given by $Trace(e^A)$. Proposition 4.3.1 remains valid for invariant power series (with the same proof).

The map assigning the cohomology class $[P(R)]$ to an invariant polynomial function P is a homomorphism of the algebra of invariant polynomial functions on $\mathcal{M}_r$ into the algebra $H^*_{dR}(M, \mathbb{K})$. It is called **the Chern-Weil homomorphism**.

4.4 Chern Classes: Definition and Basic Properties

Definition 4.4.1 Let $\pi : E \to M$ be a smooth complex vector bundle of rank r. Let P^k, $k = 0, \ldots, r$, be the invariant polynomial functions on the space $\mathcal{M}_r$ of complex $r \times r$-matrices determined by the identity

$$\det(A + tI) = P^0(A)t^r + P^1(A)t^{r-1} + \cdots + P^{r-1}(A)t + P^r(A).$$

If D is a connection on E and $R^D \in A^2(Hom(E, E))$ is the curvature tensor of D, the differential form on M of degree $2k$

$$c_k^D = P^k(\frac{i}{2\pi}R^D)$$

is called the k**th Chern form of** D.

Henceforth, we assume that the base manifold of the bundles we are considering is paracompact. Then every bundle E admits a connection D. By Proposition 4.3.1, the cohomology class of the form c_k^D does not depend on the choice of the connection D.

Definition 4.4.2 The cohomology class

$$c_k(E) = [P^k(\frac{i}{2\pi}R^D)] \in H_{dR}^{2k}(M, \mathbb{C})$$

is called the k**th Chern class**.

Example 4.4.1 Let $L \to M$ be a Hermitian holomorphic line bundle with metric h. The first Chern class $c_1(L)$ of such a bundle can be computed in the following way. Let $\{U_\alpha, \varphi_\alpha\}$ be an atlas of (holomorphic) trivializations of the bundle L with transition functions $g_{\alpha\beta} : U_\alpha \cap U_\beta \to \mathbb{C} \setminus \{0\}$. Set $e_\alpha(x) = \varphi_\alpha^{-1}(x, 1)$ for $x \in U_\alpha$ and $h_\alpha = h(e_\alpha, e_\alpha)$. As noted in Chap. 2, the locally defined curvature forms with respect to the frames e_α of the Chern connection of the bundle L glue together to a global $(1, 1)$-form Θ on M such that

$$\Theta|U_\alpha = -\partial\bar{\partial} \ln h_\alpha.$$

The first Chern class of the bundle L is the cohomology class of the form

$$Trace \frac{i}{2\pi}\Theta = \frac{i}{2\pi}\Theta;$$

on U_α, this form is equal to $\dfrac{1}{2\pi i}\partial\bar{\partial} \ln h_\alpha$. The form $\dfrac{i}{2\pi}\Theta$ is called the Chern form of the metric h.

Problem 4.4.1 *Show that the Chern form of h is real-valued.*

If R is the curvature tensor of the Chern connection on L, we have $R(X, Y)e_\alpha = \Theta(X, Y)e_\alpha$ for $X, Y \in TM$. Let J be the complex structure of the bundle L. Then

$$i\Theta(X, Y) = h(iR(X, Y)e_\alpha, e_\alpha) = h(R(X, Y)(ie_\alpha), e_\alpha) = h(R(X, Y)Je_\alpha, e_\alpha).$$

The left-hand side of this identity is a real number, hence for the Euclidean metric $g = Re\, h$, $h(R(X, Y)Je_\alpha, e_\alpha) = g(R(X, Y)Je_\alpha, e_\alpha)$. Thus, the real-valued 2-form $c_1(X, Y) = \frac{1}{2\pi} g(R(X, Y)Je_\alpha, e_\alpha)$ represents the first Chern class of L.

In addition to the above consideration, note that if $L \to M$ is a holomorphic line bundle over a Kähler manifold M and ω is a real-valued closed $(1, 1)$-form on M, whose cohomology class in $H^2_{dR}(M, \mathbb{R})$ coincides with the first Chern class $c_1(L)$ of L, the bundle L admits an Hermitian metric h whose Chern form is ω, see Voisin [22], Sect. 7.1.3, Theorem 7.10. The proof is based on the $\partial\overline{\partial}$-lemma: Let ω be a form on a Kähler manifold that is both ∂ and $\overline{\partial}$-closed. If ω is d-, or ∂-, or $\overline{\partial}$-exact, then there is a form χ such that $\omega = \partial\overline{\partial}\chi$ (Voisin [22], Sect. 6.1.3, Theorem 6.17)

The sum of all classes

$$c(E) = \sum_{k=0}^{r} c_k(E) = [\det(\frac{i}{2\pi} R^D + I)]$$

is called the **total Chern class**.

Let $e_1, ..., e_r$ be a frame of E and $\Theta = [\Theta_{ij}]$ the curvature matrix of a connection D with respect to this frame. Then, by (4.5),

$$\det(\frac{i}{2\pi} R^D + I) = \det(\frac{i}{2\pi}\Theta + I) = \sum_\sigma \epsilon(\sigma)(\frac{i}{2\pi}\Theta_{1,\sigma(1)} + \delta_{1,\sigma(1)}) \wedge \ldots \wedge (\frac{i}{2\pi}\Theta_{r,\sigma(r)} + \delta_{r,\sigma(r)}), \tag{4.10}$$

where the sum is taken over all permutations σ of the numbers $\{1, \ldots, r\}$ and $\epsilon(\sigma)$ is the signature of σ.

The form

$$ch(E) = Trace(\exp(\frac{i}{2\pi} R^D))$$

is called the **Chern character**.

If M is an almost complex manifold, the kth Chern class of the complex bundle $T^{1,0}M$ is called the kth Chern class $c_k(M)$ of the manifold M.

Proposition 4.4.1 *If $E \to M$ is a Hermitian vector bundle and D is a connection on E compatible with the metric, then the Chern forms c_k^D are real-valued on TM.*

Proof It is enough to show that the matrix representation of the Chern forms with respect to a frame consists of real-valued differential forms.

Let $e_1, \ldots, e_r$ be an orthonormal (unitary) frame of E and Θ the curvature matrix of D with respect to this frame. Then, as we have noted in Chap. 2, Sect. 2.4,

$$\Theta +^t \overline{\Theta} = 0.$$

Note also that for $X, Y \in TM$

$$\overline{\Theta}(X, Y) = \overline{\Theta(X, Y)}.$$

Therefore

$$\begin{aligned}
\det(\frac{i}{2\pi}\Theta(X, Y) + I) &= \det(-\frac{i}{2\pi}\,{}^t\overline{\Theta(X, Y)} + I) \\
&= \det {}^t(-\frac{i}{2\pi}\overline{\Theta(X, Y)} + I) \\
&= \det(-\frac{i}{2\pi}\overline{\Theta(X, Y)} + I) \\
&= \det(\overline{\frac{i}{2\pi}\Theta(X, Y))} + I),
\end{aligned}$$

where the third identity follows from (4.10) and the fact that the exterior product of 2-forms is commutative.

□

Considering the real-valued p-forms on M as complex-valued, we get a natural embedding of $H^p_{dR}(M, \mathbb{R})$ into $H^p_{dR}(M, \mathbb{C})$. Further on, $H^p_{dR}(M, \mathbb{R})$ will be regarded as a subspace of $H^p_{dR}(M, \mathbb{C})$.

Since every complex vector bundle over a paracompact manifold admits a Hermitian metric and a connection compatible with it, Proposition 4.4.1 implies the following.

Corollary 4.4.1 *The Chern classes $c_k(E) \in H^{2k}_{dR}(M, \mathbb{R})$.*

An obvious property of the Chern classes is that if $E \to M$ and $E' \to M$ are isomorphic bundles, then $c(E) = c(E')$. Other important properties are collected in the following statement.

Proposition 4.4.2 *(1) $c_0(E) = 1 \in \mathbb{R} \subset H^0_{dR}(M, \mathbb{R})$.*
(2) Let $f : M \to N$ be a smooth map, $E \to N$ a complex vector bundle, and $f^*E \to M$ its pull-back bundle. Then

$$c(f^*E) = f^*c(E), \quad (\textit{Naturality}).$$

(3) If $E' \to M$ and $E'' \to M$ are complex vector bundles,

$$c(E' \oplus E'') = c(E').c(E''), \quad (\textit{Whitney sum formula}),$$

where the product is taken in the ring $H^*_{dR}(M, \mathbb{R})$ of de Rham cohomology.

(4) If $E^* \to M$ is the dual bundle of $E \to M$, then

$$c_k(E^*) = (-1)^k c_k(E).$$

(5) If $\mathcal{O}(1)$ is the dual bundle of the tautological bundle $\mathcal{O}(-1)$ over $\mathbb{CP}^n$, then $c(\mathcal{O}(1)) = 1 + [\omega_n]$, where ω_n is the normalized Kähler form of $\mathbb{CP}^n$ (Normalization).

Proof (1) is a straightforward consequence of the definition of the Chern classes since $P^0(A) = 1$.

(2) Let D be a connection on E and $\widetilde{D}$ the induced connection on f^*E. Let $e_1, \ldots, e_r$ be a frame of E and Θ the curvature matrix of D with respect to this frame. Then $\widetilde{e_1} = e_1 \circ f, \ldots, \widetilde{e_r} = e_r \circ f$ is a frame of f^*E and if $\widetilde{\Theta}$ is the curvature matrix of $\widetilde{D}$ with respect to $\widetilde{e_1}, \ldots, \widetilde{e_r}$, then $\widetilde{\Theta} = f^*\Theta$. This implies (2).

(3) Let D' and D'' be connections on E' and E''. Denote the connection induced on $E' \oplus E''$ by $D = D' \oplus D''$. Let $e' = (e'_1, \ldots, e'_r)$ and $e'' = (e''_1, \ldots, e''_s)$ be frames of E' and E''. Denote by Θ', Θ'' and Θ the curvature matrices of D', D'' and D with respect to the frames e', e'' and $e = (e'_1, \ldots, e'_r, e''_1, \ldots, e''_s)$, respectively. Then, in the block-matrix form corresponding to the direct sum $E' \oplus E''$,

$$\Theta = \begin{bmatrix} \Theta' & 0 \\ 0 & \Theta'' \end{bmatrix}.$$

Therefore

$$\begin{aligned} \det(\frac{i}{2\pi}\Theta + I) &= \det \begin{bmatrix} \frac{i}{2\pi}\Theta' + I & 0 \\ 0 & \frac{i}{2\pi}\Theta'' + I \end{bmatrix} \\ &= \det(\frac{i}{2\pi}\Theta' + I) \wedge \det(\frac{i}{2\pi}\Theta'' + I). \end{aligned}$$

This implies (3).

(4) Let D be a connection on E and D^* the induced connection on E^*. Let $e_1, \ldots, e_r$ be a frame of E and $e_1^*, \ldots, e_r^*$ the dual frame of E^* If Θ and Θ^* are the corresponding curvature matrices of D and D^*,

$$\Theta^* = -{}^t\Theta.$$

Then

$$\det(\frac{i}{2\pi}\Theta^* + I) = \det(-\frac{i}{2\pi}{}^t\Theta + I) = \det(-\frac{i}{2\pi}\Theta + I).$$

It follows that

$$c_k^{D^*} = P^k(\frac{i}{2\pi}\Theta^*) = P^k(-\frac{i}{2\pi}\Theta) = (-1)^k P^k(\frac{i}{2\pi}\Theta) = (-1)^k c_k^D.$$

(5) Let V be a Hermitian vector space with metric h. This metric induces a metric on the trivial bundle $\underline{V} = \mathbb{P}(V) \times V$ in an obvious way. The tautological bundle

$\mathcal{O}(-1) \to \mathbb{P}(V)$ is a subbundle of $\underline{V}$ and we endow $\mathcal{O}(-1)$ with the metric induced by that of $\underline{V}$.

Let $\xi \in \mathbb{P}(V)$. Choose an orthonormal basis $v_0, \dots, v_n$ of V such that $\xi = [v_0]$. Then ξ lies in the coordinate neighbourhood

$$U_0 = \{[v_0 + z_1 v_1 + \cdots + z_n v_n] : z_k \in \mathbb{C}\}.$$

As in Chap. 3, Sect. 3.7, let $\varepsilon : U_0 \to \mathcal{O}(-1)$ be the frame of the line bundle $S = \mathcal{O}(-1)$ defined by $\varepsilon(z) = v_0 + z_1 v_1 + \cdots + z_n v_n$. Recall that the connection form of the canonical connection on S with respect to ε is

$$\theta^S = \partial h(\varepsilon, \varepsilon).h(\varepsilon, \varepsilon)^{-1} = \partial \ln h(\varepsilon, \varepsilon) = \partial \ln(1 + |z|^2) = \frac{1}{1 + |z|^2} \sum_{k=1}^{n} \overline{z}_k dz_k.$$

Hence, the curvature matrix with respect to ε is

$$\Theta^S = \overline{\partial}\theta^s = -\sum_{k,j} \frac{\delta_{kj}(1 + |z|^2) - \overline{z}_k z_j}{(1 + |z|^2)^2} dz_k \wedge d\overline{z}_j.$$

Therefore

$$\frac{i}{2\pi}\Theta^S = -\omega_n,$$

where ω_n is the normalized Kähler form of $\mathbb{P}(V) \cong \mathbb{CP}^n$. By the latter identity

$$c(S) = 1 - [\omega_n].$$

This and (4) imply

$$c(S^*) = 1 + [\omega_n].$$

□

Remark 4.4.1 Properties (1), (2), (3) and (5) uniquely determine the Chern classes (see Hirzebruch [12], Chap. I, Sect. 4.2).

It follows from Proposition 4.4.2 that $c_1(\mathcal{O}(k)) = k[\omega_n]$. This justifies the notation $\mathcal{O}(k)$.

Identity $c_k(E) = (-1)^k c_k(E^*)$ shows that a complex vector bundle may not be isomorphic to its dual. In contrast, every real vector bundle (over a paracompact manifold) is isomorphic to its dual, for example via the isomorphism defined in an obvious way by an Eulidean metric on the bundle.

Problem 4.4.2 *Let E be a complex vector bundle and $\overline{E}$ its conjugate complex bundle. Take a Hermitian metric h on E and define a map $\varphi : \overline{E} \to E^*$ by $\varphi(u)(v) = h(v, u)$ for $u \in \overline{E}$ and $v \in E$. Prove that the map φ is an isomorphism.*

Thus, by Proposition 4.4.2,

$$c_k(E) = (-1)^k c_k(\overline{E}). \tag{4.11}$$

Hence E and $\overline{E}$ are not isomorphic in the general case.

Comparing the Chern classes belonging to the same cohomology group on both sides of the identity $c(E' \oplus E'') = c(E').c(E'')$, we get

$$c_k(E' \oplus E'') = \sum_{r+s=k} c_r(E').c_s(E'').$$

In the general case, computing the Chern classes of a tensor, exterior or symmetric product of complex vector bundles is not an easy task. Often, for this purpose, the above formula and the following splitting principle are used so that the computation is reduced to the computation of the Chern class of line bundles.

Proposition 4.4.3 *Let $\pi : E \to M$ be a smooth complex vector bundle over a paracompact manifold M. Then there is a manifold $F(E)$ and a smooth map $\sigma : F(E) \to M$ such that*

(1) the pull-back bundle $\sigma^* E \to F(E)$ of the bundle $E \to M$ decomposes into the direct sum of a finite number of complex line bundles

$$\sigma^* E = L_1 \oplus \ldots \oplus L_n;$$

(2) the homomorphism $H^*_{dR}(M, \mathbb{C}) \to H^*_{dR}(F(E), \mathbb{C})$ induced by the map $\sigma^* : A^*(M) \to A^*(F(E), \omega \to \sigma^* \omega$, is injective.

This proposition holds also in the category of holomorphic vector bundles over paracompact complex manifolds.

A proof and applications of the splitting principle can be seen in Bott and Tu [2], Chap. IV, § 21.

Proposition 4.4.4 *Let $E \to M$ be a smooth complex vector bundle of rank r.*

(1) If E is the trivial bundle, then $c_k(E) = 0$ for $k \geq 1$, i.e., $c(E) = 1$.

(2) If $E \cong E' \oplus T_s$, where T_s is the trivial vector bundle of rank s, then $c_j(E) = c_j(E')$ for $j = 1, \ldots, r - s$ and $c_k(E) = 0$ for $k = r - s + 1, \ldots, r$, i.e., $c(E) = c(E')$.

Proof (1) The curvature of the trivial connection d on the trivial bundle vanishes and (1) is an obvious consequence of the definition of the Chern classes.
(2) By Proposition 4.4.2 item (3),

$$c(E) = c(E').c(T_s) = c(E').$$

Moreover, E' is of rank $r - s$, hence

$$1 + c_1(E) + \cdots + c_r(E) = 1 + c_1(E') + \cdots + c_{r-s}(E').$$

Comparing the Chern classes belonging to the same cohomology group in both sides, we get (2). □

Example 4.4.2 The total Chern class of the complex projective space.

Let $Q = \underline{\mathbb{C}}^{n+1}/\mathcal{O}(-1)$ be the quotient bundle of the trivial bundle $\underline{\mathbb{C}}^{n+1} = \mathbb{CP}^n \times \mathbb{C}^{n+1}$ by the tautological bundle $\mathcal{O}(-1)$ over $\mathbb{CP}^n$. Then we have the following exact sequence of bundles

$$0 \to \mathcal{O}(-1) \to \underline{\mathbb{C}}^{n+1} \to Q \to 0,$$

where the second map is the inclusion and the third one is the natural projection. Tensoring with $\mathcal{O}(1)$, we get the exact sequence

$$0 \to \underline{\mathbb{C}} \to \oplus^{n+1}\mathcal{O}(1) \to Q \otimes \mathcal{O}(1) \to 0.$$

The third member of this sequence is obtained by the trivial observation that $\mathbb{C}^{n+1} \otimes V = (\mathbb{C} \oplus \ldots \oplus \mathbb{C}) \otimes V \cong (\mathbb{C} \otimes V) \oplus \ldots \oplus (\mathbb{C} \otimes V) \cong V \oplus \ldots \oplus V = \oplus^{n+1} V$ for any complex vector space V.

In Chap. 3, we have proved that $Q \otimes \mathcal{O}(1) \cong T^h\mathbb{CP}^n$. Thus, we obtain the following exact sequence

$$0 \to \underline{\mathbb{C}} \to \oplus^{n+1}\mathcal{O}(1) \to T^h\mathbb{CP}^n \to 0.$$

This sequence is called the Euler sequence (for another proof of the exactness of the Euler sequence, see Griffiths and Harris [9], Chap. 3, Sect. 3).

The exactness of the Euler sequence implies $\oplus^{n+1}\mathcal{O}(1) \cong T^h\mathbb{CP}^n \oplus \underline{\mathbb{C}}$ as smooth bundles. Hence, by Proposition 4.4.4,

$$c(\oplus^{n+1}\mathcal{O}(1)) = c(T^h\mathbb{CP}^n).$$

Furthermore, by Proposition 4.4.2,

$$c(\oplus^{n+1}\mathcal{O}(1)) = c(\mathcal{O}(1))^{n+1} = (1 + [\omega_n])^{n+1}.$$

Therefore

$$c(T^h\mathbb{CP}^n) = (1 + [\omega_n])^{n+1} = \sum_{k=0}^{n+1} \binom{n+1}{k} [\omega_n]^k,$$

where the last equality holds because the exterior product of differential forms of even degrees is commutative. Note that $\omega_n^{n+1} = 0$ since this form is of degree $2n+2$, while the real dimension of $\mathbb{CP}^n$ is $2n$.

It follows from the formula for the total Chern class $c(T^h\mathbb{CP}^n)$ we derived that

$$c_1(T^h\mathbb{CP}^n) = (n+1)[\omega_n].$$

Proposition 4.4.5 *Let $E \to M$ be a complex vector bundle of rank r and L a line bundle over the manifold M. Then*

$$c_1(E \otimes L) = c_1(E) + r c_1(L).$$

Proof Let D^E and D^L be connections on E and L. Denote by $D^{E\otimes L}$ the connection on $E\otimes L$ induced by D^E and D^L. Let $e_1, \ldots, e_r$ be a frame of E and f a frame of L. If Θ^E and Θ^L are the curvature matrices of D^E and D^L with respect to these frames, then the curvature matrix of $D^{E\otimes L}$ with respect to the frame $e_1 \otimes f, \ldots, e_r \otimes f$ is

$$\Theta^{E\otimes L} = \Theta^E + \Theta^L I_r,$$

where I_r is the unit $r \times r$-matrix; here Θ^E is a $r \times r$-matrix of 2-forms and Θ^L is a 2-form. Therefore

$$c_1(E \otimes L) = Trace \frac{i}{2\pi}\Theta^{E\otimes L} = c_1(E) + rc_1(L).$$

□

4.5 Holomorphic Line Bundle Defined by a Complex Hypersurface. Computation of the Chern Classes of Hypersurfaces

Now, we show that every closed hypersurface in a complex manifold M defines a holomorphic line bundle over M. Recall that, by definition, a hypersurface in M is a complex submanifold of M of codimension 1 whose topology coincides with that induced by the topology of M.

Let Σ be a topological subspace of M. Using the implicit function theorem, one can prove that the following three conditions are equivalent (if necessary, see Griffiths and Harris [9], Chap. 0, Sect. 2 or Narasimhan [20], Chap. 2, § 2.5)

(1) Σ has the structure of a hypersurface of M.
(2) For every point $p \in M$, there is a chart $(U, z_1, \ldots, z_n)$ of M at p such that

$$U \cap \Sigma = \{x \in U : z_1(x) = 0\}.$$

(3) For every point $p \in M$, there exist a neighbourhood U of p and a holomorphic function f in U such that

$$U \cap \Sigma = \{x \in U : f(x) = 0\} \text{ and } (df)_x \neq 0 \text{ for every } x \in U \cap \Sigma.$$

A function f with the property (3) is said to be a defining function of Σ in U.

Let Σ be a hypersurface in M and $(U, z_1, \ldots, z_n)$ a chart of M with the property (2). Then the functions $u_k = z_k|(U \cap \Sigma),\ k = 2, \ldots, n$, constitute a local coordinate system of Σ. Moreover, it follows from (2) that if a topological subspace of M admits the structure of a hypersurface, then this structure is uniquely determined.

To construct a line bundle when a hypersurface is given, we need the following lemma.

Lemma 4.5.1 *Let G be an open set in $\mathbb{C}^n$ intersecting the hypersurface $\Sigma = \{z_1 = 0\}$. Let f and g be holomorphic functions on G vanishing only on $G \cap \Sigma$ and such that $df \neq 0$ and $dg \neq 0$ at the points of this set. Then the function $\frac{f}{g}$ can be uniquely extended to a non-vanishing holomorphic function on G.*

Proof Let $a = (a_1, a_2, \ldots, a_n) \in G \cap \Sigma$ and let $D = \{z \in G : |z_k - a_k| < r,\ 1 \leq k \leq n\}$ be a polydisc with centre at the point a (a product of discs) lying in G; here $a_1 = 0$. The function f in the polydisc D is uniquely represented as the sum of a power series, its Taylor series:

$$f(z) = \sum_{\alpha_1, \alpha_2, \ldots, \alpha_n \geq 0}^{\infty} c_{\alpha_1 \alpha_2 \ldots \alpha_n} z_1^{\alpha_1} (z_2 - a_2)^{\alpha_2} \ldots (z_n - a_n)^{\alpha_n}.$$

In this series, $c_{0\alpha_2 \ldots \alpha_n} = 0$ since

$$0 = f(0, z_2, \ldots, z_n) = \sum_{\alpha_2, \ldots, \alpha_n \geq 0}^{\infty} c_{0, \alpha_2 \ldots \alpha_n} (z_2 - a_2)^{\alpha_2} \ldots (z_n - a_n)^{\alpha_n}$$

for every $(z_2, \ldots, z_n)$ with $|z_j - a_j| < r,\ 2 \leq j \leq n$. Hence, the power series expansion of f in D is

$$f(z) = c_{10\ldots0} z_1 + \sum_{\alpha_1 \geq 1, \alpha_2, \ldots, \alpha_n}^{\infty} c_{\alpha_1 \alpha_2 \ldots \alpha_n} z_1^{\alpha_1} (z_2 - a_2)^{\alpha_2} \ldots (z_n - a_n)^{\alpha_n},$$

where the sum is taken over $(\alpha_1, \alpha_2, \ldots, \alpha_n) \neq (1, 0, \ldots, 0)$. Therefore for $z \in D \backslash \Sigma$,

$$\frac{f(z)}{z_1} = c_{10\ldots0} + \sum_{\alpha_1 \geq 1, \alpha_2, \ldots, \alpha_n}^{\infty} c_{\alpha_1 \alpha_2 \ldots \alpha_n} z_1^{\alpha_1 - 1} (z_2 - a_2)^{\alpha_2} \ldots (z_n - a_n)^{\alpha_n}.$$

Let $b = (b_1, \ldots, b_n)$ be a point of D with $b_k \neq a_k$, $k = 1, \ldots, n$, in particular, $b_1 \neq a_1 = 0$. The power series in the right-hand side of the latter identity is clearly convergent at the point b. By the Abel lemma, it is convergent in the polydisc $D' = \{z \in G^n : |z_k - a_k| < |b_k - a_k|,\ k = 1, \ldots, n\} \subset D$. Its sum φ is a holomorphic function in D'. By assumption, $\left(\frac{\partial f}{\partial z_1}(a), \ldots, \frac{\partial f}{\partial z_n}(a)\right) = (df)_a \neq 0$; moreover, $f(0, z_2, \ldots, z_n) = 0$. Hence, $\frac{\partial f}{\partial z_j}(a) = 0$ for $j = 2, \ldots, n$. Thus, $\varphi(a) = c_{10\ldots0} = \frac{\partial f}{\partial z_1}(a) \neq 0$. Therefore the function φ does not vanish in a neighbourhood of the point a.

Similarly, the function $\dfrac{g(z)}{z_1}$ can be extended to a non-vanishing holomorphic function ψ in a neighbourhood of the point a.

In this way, we see that for every $a \in G \cap \Sigma$ there is a neighbourhood $U_a \subset G$ of the point a such that the function

$$\frac{f(z)}{g(z)} = \frac{\dfrac{f(z)}{z_1}}{\dfrac{g(z)}{z_1}}, \quad z \in G\backslash\Sigma,$$

can be extended to a non-vanishing holomorphic function $h_a = \dfrac{\varphi}{\psi}$ on U_a. Now, set

$$h(z) = \begin{cases} \dfrac{f(z)}{g(z)} \text{ if } z \in G\backslash\Sigma \\ h_a(z) \text{ if } z \in U_a. \end{cases} \tag{4.12}$$

Suppose $U_a \cap U_b \neq \emptyset$, $a, b \in G \cap \Sigma$. Then on $(U_a \cap U_b)\backslash\Sigma$,

$$h_a = \frac{f}{g} = h_b.$$

The set $(U_a \cap U_b)\backslash\Sigma$ is dense in $U_a \cap U_b$, hence, by continuity, $h_a = h_b$ everywhere on $U_a \cap U_b$. It follows that formula (4.12) correctly defines a function on G. It is clear that this function possesses the desired properties.

The uniqueness of the extension follows from the fact that the set $G\backslash\Sigma$ is dense in G.

□

Now, let M be a complex manifold and Σ a closed hypersurface in M. For every $a \in \Sigma$, take a defining function f_a of Σ in a neighbourhood U_a of the point a. If $a \notin \Sigma$, there is a neighbourhood U_a of a such that $U_a \cap \Sigma = \emptyset$. In this case, set $f_a \equiv 1$. Thus, we obtain a family $\{U_a, f_a\}$ consisting of an open cover $\{U_a\}$ of M and holomorphic functions $\{f_a\}$ in U_a such that

$$\begin{aligned} &* \text{ If } U_a \cap \Sigma \neq \emptyset, \ f_a \text{ is a defining function of } \Sigma \text{ in } U_a; \\ &* \text{ If } U_a \cap \Sigma = \emptyset, \ f_a \equiv 1 \text{ on } U_a. \end{aligned} \tag{4.13}$$

Suppose $U_a \cap U_b \neq \emptyset$. If Σ intersects $U_a \cap U_b$, then it follows from Lemma 4.5.1 that the function $\dfrac{f_a}{f_b}$ on $(U_a \cap U_b)\backslash\Sigma$ extends to a unique non-vanishing holomorphic function g_{ab} on the whole set $U_a \cap U_b$. If Σ does not intersect $U_a \cap U_b$, the holomorphic functions f_a and f_b do not vanish on $U_a \cap U_b$, hence $g_{ab} = \dfrac{f_a}{f_b}$ is a non-vanishing holomorphic function on $U_a \cap U_b$. It is obvious that if $U_a \cap U_b \cap U_c \neq \emptyset$, then

$$g_{ab}\, g_{bc}\, g_{ca} = 1 \text{ on } (U_a \cap U_b \cap U_c)\backslash\Sigma,$$

hence this identity holds on the whole set $U_a \cap U_b \cap U_c$. Thus, the functions g_{ab} satisfy the cocycle condition. By Chap. 1, Proposition 1.3.1, these functions determine a holomorphic line bundle L over M, unique up to isomorphism, with transition functions g_{ab}. In fact, this bundle does not depend on the choice of the family $\{U_a, f_a\}$ with property (4.13). For, let $\{V_i\}$ be an open cover of M and φ_i holomorphic functions on V_i satisfying (4.13). Then the function $\frac{\varphi_i}{\varphi_j}$ on $(V_i \cap V_j)\backslash\Sigma$ extends to a non-vanishing holomorphic function h_{ij} on $V_i \cap V_j$. Denote by L' the bundle with transition functions h_{ij}. If $U_a \cap V_i \neq \emptyset$, the function $\frac{\varphi_i}{f_a}$ on $(U_a \cap V_i)\backslash\Sigma$ extends to a non-vanishing holomorphic function u_{ia} on $U_a \cap V_i$. Clearly, if $x \in U_a \cap U_b \cap V_i \cap V_j$, then

$$u_{jb}(x) = h_{ji}(x)u_{ia}(x)g_{ab}(x).$$

Therefore the functions $\{u_{ia}\}$ determine a morphism $u : L \to L'$. Since $u_{ia}(x) \neq 0$, multiplication by $u_{ia}(x)$ defines an isomorphism of the vector space $\mathbb{C}$. It follows that the bundles L and L' are isomorphic (see Chap. 1, Sect. 1.3, identity (1.4) and the discussion after it). Since $f_a = g_{ab}f_b$, the functions $\{f_a\}$ determine a holomorphic section s of L whose zero locus is exactly the hypersurface Σ. Similarly, the functions $\{\varphi_i\}$ define a section σ of L'. Clearly, $\sigma_i(x) = u_{ia}(x)f_a(x)$ for $x \in U_a \cap V_i$. Hence, the isomorphism u of L onto L' sends the section s of L into the section σ of L', $\sigma = u \circ s$. Therefore we can talk about a line bundle and its section associated to Σ without specifying which particular bundle we have in mind. The associated bundle will be denoted by $[\Sigma]$.

Let M be a complex manifold and Σ a closed hypersurface in M. Recall that the normal bundle of Σ is the quotient-bundle $N_\Sigma = (T^hM|\Sigma)/T^h\Sigma$. It is a holomorphic line bundle over Σ. Its dual bundle N^*_Σ is called the conormal bundle of Σ. The conormal bundle can be identified with the subbundle of $(T^hM|\Sigma)^*$ consisting of 1-forms (i.e., covectors) $T^h_aM \to \mathbb{C}$, $a \in \Sigma$, vanishing on the subspace $T^h_a\Sigma$ of T^h_aM. Let the associated line bundle $[\Sigma]$ be defined by the family $\{U_a, f_a\}$. If $U_a \cap \Sigma \neq \emptyset$, then $f_a \equiv 0$ on $U_a \cap \Sigma$, hence $df_a|T^h\Sigma = 0$. Therefore $df_a|(U_a \cap \Sigma) : U_a \cap \Sigma \to (T^hM|\Sigma)^*$ is a local section of N^*_Σ. Furthermore,, for every $x \in U_a \cap \Sigma$, $(df_a)_x \neq 0$, hence df_a is a frame of N^*_Σ on $U_a \cap \Sigma$. Let g_{ab} be the transition functions of $[\Sigma]$ defined by the family $\{U_a, f_a\}$. Then, on $U_a \cap U_b \cap \Sigma$,

$$df_a = d(g_{ab}f_b) = dg_{ab} \cdot f_b + g_{ab}df_b = g_{ab}df_b.$$

The transition function g^*_{ab} of the bundle N^*_Σ is the transition function from the frame df_b to the frame df_a. It follows that g_{ba} are transition functions of the bundle N^*_Σ. Therefore $N^*_\Sigma \cong [\Sigma]^*|\Sigma$. In this way, we obtain

$$N_\Sigma \cong [\Sigma]|\Sigma.$$

Example 4.5.1 Let Σ be the hypersurface in $\mathbb{CP}^n$ defined in homogeneous coordinates by

$$z_0 + \cdots + z_n = 0.$$

On the coordinate neighbourhood $U_i = \{[z] : z_i \neq 0\}$, Σ can be defined by the function

$$f_i([z]) = \sum_{k=0}^{n} \frac{z_k}{z_i} = 1 + \xi_0 + \cdots + \xi_{i-1} + \xi_{i+1} + \cdots + \xi_n,$$

where $\xi_k = \dfrac{z_k}{z_i}$, $k \neq i$, are the standard coordinates on U_i. On $U_i \cap U_j$, we have

$$\frac{f_i}{f_j} = \frac{z_j}{z_i}.$$

Therefore the transition functions of the bundle $[\Sigma]$ are $\dfrac{z_j}{z_i}$. These are also the transition functions of the bundle $\mathcal{O}(1)$ for the cover $\{U_i\}$ of $\mathbb{CP}^n$. It follows that $[\Sigma] \cong \mathcal{O}(1)$. Because of this isomorphism, $\mathcal{O}(1)$ is called a **hyperplane** bundle.

Example 4.5.2 Let $r \geq 1$ be a natural number. Set

$$\Sigma = \{[z] \in \mathbb{CP}^n : z_0^r + \cdots + z_n^r = 0\}.$$

On the set U_i, Σ is given as the zero locus of the function

$$f_i([z]) = \sum_{k=0}^{n} \left(\frac{z_k}{z_i}\right)^r = 1 + \xi_1^r + \cdots + \xi_{i-1}^r + \xi_{i+1}^r + \cdots + \xi_n^r.$$

Clearly, $df_i = \sum_{j \neq i} r\xi_j^{r-1} d\xi_j$. Since $1 + \sum_{j \neq i} \xi_j^r = 0$ on Σ, $\xi_j \neq 0$ for at least one j. Hence, $df_i \neq 0$ at the points of Σ. Therefore f_i is a defining function of Σ in U_i. On $U_i \cap U_j$,

$$\frac{f_i}{f_j} = \left(\frac{z_j}{z_i}\right)^r.$$

This shows that $\left(\dfrac{z_j}{z_i}\right)^r$ are transition functions of the bundle $[\Sigma]$. This functions are also transition functions of the bundle $\otimes^r \mathcal{O}(1)$. Therefore

$$[\Sigma] \cong \otimes^r \mathcal{O}(1) = \mathcal{O}(r).$$

In order to compute the Chern classes of the manifold Σ, we use the decomposition

$$T^h\mathbb{CP}^n|\Sigma \cong T^h\Sigma \oplus N_\Sigma,$$

where $N_\Sigma \cong [\Sigma]|\Sigma \cong \otimes^r \mathcal{O}(1)|\Sigma$.

Let $\iota : \Sigma \hookrightarrow \mathbb{CP}^n$ be the inclusion map. Then $T^h\mathbb{CP}^n|\Sigma = \iota^*(T^h\mathbb{CP}^n)$, the pull-back bundle. Hence, the total Chern class c of the bundle $T^h\mathbb{CP}^n|\Sigma$ is

$$c = \iota^* c(\mathbb{CP}^n).$$

We have computed in Example 4.4.2 that

$$c(\mathbb{P}^n) = (1 + [\omega_n])^{n+1}.$$

Set $\eta = \iota^*[\omega_n]$. Then $c = (1+\eta)^{n+1}$, where, by Proposition 4.4.2,

$$(1+\eta)^{n+1} = c(\Sigma).c(N_\Sigma).$$

As we have seen above

$$N_\Sigma \cong [\Sigma]|\Sigma \cong \otimes^r \mathcal{O}(1)|\Sigma.$$

By Proposition 4.4.2,

$$c(\mathcal{O}(1)) = 1 + [\omega_n].$$

Applying Proposition 4.4.5, we get

$$c(\otimes^r \mathcal{O}(1)) = 1 + r[\omega_n].$$

Hence,

$$c(\otimes^r \mathcal{O}(1)|\Sigma) = 1 + r\eta.$$

Thus, we have the identity

$$(1+\eta)^{n+1} = (1 + c_1(\Sigma) + \cdots + c_{n-1}(\Sigma)).(1 + r\eta).$$

Set for brevity

$$c_i = c_i(\Sigma).$$

Then

$$\sum_{k=0}^{n+1} \binom{n+1}{k} \eta^k = 1 + c_1 + r\eta + (c_2 + rc_1\eta) + (c_3 + rc_2\eta) + \cdots + (c_{n-1} + rc_{n-2}\eta) + rc_{n-1}\eta.$$

It follows

$$\begin{aligned}
c_1 + r\eta &= (n+1)\eta \\
c_2 + rc_1\eta &= \tbinom{n+1}{2}\eta^2 \\
c_3 + rc_2\eta &= \tbinom{n+1}{3}\eta^3 \\
&\cdots\cdots\cdots\cdots \\
c_{n-1} + rc_{n-2}\eta &= \tbinom{n+1}{n-1}\eta^{n-1}.
\end{aligned}$$

From these identities, we successively obtain

$$c_1 = (n+1-r)\eta$$
$$c_2 = \left(\binom{n+1}{2} - r(n+1-r)\right)\eta^2$$
$$c_3 = \left(\binom{n+1}{3} - r\binom{n+1}{2} + r^2(n+1-r)\right)\eta^3$$
$$\dots\dots\dots\dots\dots\dots\dots\dots\dots\dots$$
$$c_{n-1} = \eta^{n-1}\sum_{j=0}^{n-1}(-1)^{n-1-j}\binom{n+1}{j}r^{n-1-j}.$$

Therefore

$$c_k = \eta^k \sum_{j=0}^{k}(-1)^{k-j}\binom{n+1}{j}r^{k-j}, \quad k = 1, \dots, n-1.$$

4.6 Pontrjagin Classes of a Real Vector Bundle

Let $E \to M$ be a real vector bundle of rank r. The k**th Pontrjagin class** of E is defined as

$$p_k(E) = (-1)^k c_{2k}(E^{\mathbb{C}}) \in H^{4k}_{dR}(M, \mathbb{R}), \quad k = 1, \dots, r,$$

where $c_{2k}(E^{\mathbb{C}})$ is the $2k$th Chern class of the complexification $E^{\mathbb{C}}$ of E (for the latter see Chap. 1, Sect. 1.3, if necessary).

Remark 4.6.1 As we have remarked in Chap. 1, Sect. 1.4 (the comments after Problem 1.4.6) the complex vector bundles $E^{\mathbb{C}}$ and $\overline{E^{\mathbb{C}}}$ are isomorphic (but, recall that, in general, a complex vector bundle may not be isomorphic to its conjugate). By (4.11), $c_m(E^{\mathbb{C}}) = (-1)^m c_m(\overline{E^{\mathbb{C}}}) = (-1)^m c_m(E^{\mathbb{C}})$. Therefore if m is odd, $c_m(E^{\mathbb{C}}) = 0$ in the de Rham cohomology.

By definition, $p_k(E) = (-1)^k P^{2k}(\frac{i}{2\pi}R^D)$ where D is any connection on the bundle $E^{\mathbb{C}}$ and $P^{2k}(\frac{i}{2\pi}R^D)$ is the $2kth$ elementary symmetric function of the eigenvalues of $\frac{i}{2\pi}R^D$. Then $p_k(E) = (-1)^k i^{2k}[P^{2k}(\frac{1}{2\pi}R^D)] = [P^{2k}(\frac{1}{2\pi}R^D)]$. We can take any connection ∇ on the bundle E and extend it to a connection D on the bundle $E^{\mathbb{C}}$ in an obvious way: $D_X(s_1 + is_2) = \nabla_X s_1 + i\nabla_X s_2$. If $e_1, \dots, e_r$ is a frame of the real bundle E, consider each e_i as a section $x \to (e_i(x), 0)$ of $E^{\mathbb{C}} = E \oplus E$. Then $e_1, \dots, e_r$ is a frame of the complex bundle $E^{\mathbb{C}}$ and the corresponding curvature matrices of ∇ and D coincide. Therefore

$$p_k(E) = [P^{2k}(\frac{1}{2\pi}R^{\nabla})] = (\frac{1}{2\pi})^{2k}[P^{2k}(R^{\nabla})].$$

Moreover, if m is an odd number, $0 = c_m(E^{\mathbb{C}}) = [P^m(\frac{i}{2\pi}R^\nabla)] = i^m[P^m(\frac{1}{2\pi}R^\nabla)]$. Hence,

$$[P^m(\frac{1}{2\pi}R^\nabla)] = 0, \quad m \ odd.$$

It follows that

$$[det(\frac{1}{2\pi}R^\nabla + I)] = \sum_{j=0}^{r}[P^j(\frac{1}{2\pi}R^\nabla)] = p_0(E) + p_1(E) + \cdots + p_{[r/2]}(E) \quad (4.14)$$

The sum

$$p(E) = p_0(E) + p_1(E) + \cdots + p_{[r/2]}(E) \in H^*_{dR}(M, \mathbb{R}).$$

is called the **total Pontrjagin class** of E. The form $p_k^\nabla = (\frac{1}{2\pi})^{2k} P^{2k}(R^\nabla)$ is called the kth Pontrjagin form of ∇.

Proposition 4.6.1 *Let E be a real vector bundle*

(1) $p_0(E) = 1 \in \mathbb{R} \subset H^0_{dR}(M, \mathbb{R})$.

(2) Let $f : M \to N$ be a smooth map, $E \to N$ a real vector bundle, and $f^*E \to M$ its pull-back bundle. Then

$$p(f^*E) = f^*p(E).$$

(3) If $E' \to M$ and $E'' \to M$ are real vector bundles,

$$p(E' \oplus E'') = p(E').p(E''),$$

where the product is taken in the ring $H^*_{dR}(M, \mathbb{R})$ of de Rham cohomology.

(4) If $E^* \to M$ is the dual bundle of $E \to M$, then

$$p_k(E^*) = p_k(E).$$

Proof The claims (1) and (2) follow from the corresponding properties of the Chern numbers (recall that $(f^*E)^{\mathbb{C}} \cong f^*E^{\mathbb{C}}$). Also,

$$p_k(E^*) = (-1)^k c_{2k}((E^*)^{\mathbb{C}}) = (-1)^k c_{2k}((E^{\mathbb{C}})^*) = (-1)^k(-1)^{2k} c_{2k}(E^{\mathbb{C}}) = p_k(E)$$

and (4) follows. Using (4.14), the proof of (3) is similar to the proof of the Whitney sum formula for the Chern classes in Proposition 4.4.2. □

Now, let E be a complex vector bundle of rank r and $E^{1,0}$, $E^{0,1}$ the subbundles of $(1, 0)$ and $(0, 1)$ vectors of $E^{\mathbb{C}}$ (Chap. 1, Example 1.7.2). The complex bundle $E^{1,0}$ is isomorphic to E, while $E^{0,1}$ is isomorphic to the conjugate bundle $\overline{E}$. Thus, $E^{\mathbb{C}} = E \oplus \overline{E}$ and

$$\sum_{k=0}^{r}(-1)^k p_k(E) = \sum_{k=0}^{r} c_{2k}(E^{\mathbb{C}}) = \sum_{i=0}^{2r} c_i(E^{\mathbb{C}}) = c(E^{\mathbb{C}}) = c(E \oplus \overline{E}) = c(E).c(\overline{E})$$
$$= \sum_{i=0}^{r} c_i(E). \sum_{j=0}^{r}(-1)^j c_j(E).$$

Comparing the classes belonging to the same cohomology group on both sides of this identity, one can express the Pontrjagin classes of a complex vector bundle in terms of its Chern classes. For example,

$$p_1(E) = c_1(E)^2 - 2c_2(E), \quad p_2(E) = c_2(E)^2 - 2c_1(E)c_3(E) + 2c_4(E)$$
$$p_3(E) = -c_3(E)^2 + 2c_2(E)c_4(E) + 2c_6(E).$$

Example 4.6.1 The Pontrjagin classes of a smooth manifold are that of its tangent bundle. In order to compute the Pontrjagin classes of the unit sphere, we note that if $i : S^n \hookrightarrow \mathbb{R}^{n+1}$ is the standard embedding and N is the normal bundle of the sphere, $i^*T\mathbb{R}^{n+1} = TS^n \oplus N$. Passing to the complexified bundles, we clearly get $(i^*T\mathbb{R}^{n+1})^{\mathbb{C}} = T^{\mathbb{C}}S^n \oplus N^{\mathbb{C}}$. The bundles $i^*T\mathbb{R}^{n+1}$ and N are trivial, hence their complexified bundles are also trivial. It follows that

$$1 = c((i^*T\mathbb{R}^{n+1})^{\mathbb{C}}) = c(T^{\mathbb{C}}S^n \oplus N^{\mathbb{C}}) = c(T^{\mathbb{C}}S^n).c(N^{\mathbb{C}}) = c(T^{\mathbb{C}}S^n).$$

Therefore $p_0(S^n) = 1$ and $p_k(S^n) = 0$ for $k \geq 1$.

4.7 Orientation of a Vector Bundle

Recall first the notion of an orientation on a vector space. We say that two ordered bases $e = (e_1, \ldots, e_n)$ and $e' = (e'_1, \ldots, e'_n)$ of a n-dimensional real vector space V determine the same orientation if the transition matrix from e to e' has positive determinant. This defines an equivalence relation on the set of ordered bases of V. Clearly, there are only two equivalence classes; if $(e_1, e_2, \ldots, e_n)$ represents one of them, $(-e_1, e_2, \ldots, e_n)$ represents the other one. An **orientation** of V is just a choice of one of these classes. If an orientation ν is fixed, the other one is called the **opposite orientation** and is usually denoted by $-\nu$. A basis representing a given orientation is called **(positively) oriented**.

If $\alpha = (\alpha_1, \ldots, \alpha_n)$ is the dual basis of a basis $e = (e_1, \ldots, e_n)$, the n-form $\omega_e = \alpha_1 \wedge \ldots \wedge \alpha_n$ is non-zero and skew-symmetric. The space $\Lambda^n V^*$ of skew-symmetric n-forms is one dimensional, so $\Lambda^n V^* \setminus \{0\}$ has two connected components. Every two forms belonging to the same component differ by a positive constant. A choice of an orientation of V is equivalent to a choice of a component of $\Lambda^n V^* \setminus \{0\}$, i.e., a non-zero n-form. Indeed, if the bases e and e' determine the same orientation and if $T = T_{e,e'}$ is the transition matrix from e to e', $\omega_{e'} = det[(T^t)^{-1}]\omega_e = (det[T])^{-1}\omega_e$, hence $\omega_{e'}$ and ω_e lie in the same component of $\Lambda^n V^* \setminus \{0\}$. Conversely, for every $\omega \in \Lambda^n V^* \setminus \{0\}$ there is an ordered basis $e = (e_1, \ldots, e_n)$ such that $\omega(e_1, \ldots, e_n) > 0$. The orientation determined by e does not depend on its choice.

For, if $e' = (e'_1, \dots, e'_n)$ is an ordered basis with the property that $\omega(e'_1, \dots e'_n) > 0$, the identity $\omega(e'_1, \dots e'_n) = det[T_{e,e'}]\omega(e_1, \dots, e_n)$ implies that e and e' determine the same orientation. Clearly, any n-form $\sigma = \lambda\omega$ with $\lambda > 0$ yields the same orientation as ω.

Similarly, if e and e' are bases of V representing an orientation, the corresponding n-vectors $e_1 \wedge \dots \wedge e_n$ and $e'_1 \wedge \dots \wedge e'_n$ lie in the same component of $\Lambda^n V \setminus \{0\}$ since they differ by a positive factor. Conversely, if $u \in \Lambda^n V \setminus \{0\}$, take an ordered basis $e = (e_1, \dots, e_n)$ of V such that $e_1 \wedge \dots \wedge e_n = \lambda u$ where $\lambda > 0$. If e' is another basis with this property, e and e' determine the same orientation. This orientation depends only on the choice of a component of $\Lambda^n V \setminus \{0\}$.

Example 4.7.1 Let V be a n-dimensional complex vector space. If $e = (e_1, \dots, e_n)$ is a complex basis of V, $a = (e_1, \dots, e_n, e_{n+1} = ie_1, \dots, e_{2n} = ie_n)$ is a basis of the real vector space V. Let T be the transition matrix from e to another complex basis $e' = (e'_1, \dots, e'_n)$ and let $T = A + iB$ where A and B are real $n \times n$-matrices. Then the transition matrix from the real basis a to the real basis $a' = (e'_1, \dots, e'_n, e'_{n+1} = ie'_1, \dots, e'_{2n} = ie'_n)$ has the following block-matrix form

$$\begin{bmatrix} A & B \\ -B & A \end{bmatrix}. \tag{4.15}$$

The determinant of this matrix is:

$$det\begin{bmatrix} A & B \\ -B & A \end{bmatrix} = det\begin{bmatrix} A+iB & B \\ -B+iA & A \end{bmatrix} = det\begin{bmatrix} A+iB & B \\ -B+iA-i(A+iB) & A-iB \end{bmatrix}$$
$$= det\begin{bmatrix} A+iB & B \\ 0 & A-iB \end{bmatrix} = det[A+iB]det[A-iB] = |det(A+iB)|^2 > 0.$$

Therefore all complex bases of V yield the same orientation of the real vector space V. This orientation is called the **canonical orientation** of V.

Remark 4.7.1 Another orientation of a complex vector space used in the literature is that determined by the basis $(e_1, ie_1, \dots, e_n, ie_n)$, where, as above, $(e_1, \dots, e_n)$ is a complex basis. The two orientations differ by $(-1)^{[n/2]}$.

Remark 4.7.2 The real representation (4.15) of a complex matrix $A + iB$ is chosen to be consistent with our preceding considerations. It coincides with that used in Kobayahsi and Nomizu, Chap. IX, Sect. 1 and is the transpose of the one in Walschap [23], Chap. 6, Sect. 8, formula (8.1).

Problem 4.7.1 *Let V be an oriented vector space. Show that the orientation on the dual space V^* defined by the dual basis of an oriented basis $e = (e_1, \dots, e_n)$ of V does not depend on the choice of e.*

An orientation of a real vector bundle $E \to M$ is a map assigning to every point $x \in M$ an orientation $\nu(x)$ of the fibre E_x with the property that for every point

$p \in M$ there is a frame $s_1, \ldots, s_r$ of sections of E in a neighbourhood of p such that for every x the basis $(s_1(x), \ldots, s_r(x))$ of the fibre E_x defines the orientation $\nu(x)$ of E_x. Loosely speaking, the latter condition means that the orientation $\nu(x)$ smoothly depends on the point x. A frame $s_1, \ldots, s_r$ yielding the orientations of the fibres is called oriented. A bundle is called **orientable** if it admits an orientation. A manifold is called oriented (orientable) if so is its tangent bundle.

Problem 4.7.2 *Prove that if the base manifold M of an orientable bundle $E \to M$ is connected, the bundle E admits exactly two orientations.*

A simple useful consequence of this is the following fact whose proof is left to the reader.

Problem 4.7.3 *Suppose that E is an orientable bundle. Then if $\xi_1, \ldots, \xi_r$ and $\eta_1, \ldots, \eta_r$ are any frames of E defined on connected open sets U and V with $U \cap V \neq \emptyset$, the determinant of the transition matrix from one of these frames to another has a constant sign on $U \cap V$.*

Problem 4.7.4 *Prove that a real bundle E of rank r is orientable if and only if its structure group can be reduced to the group $SO(r)$, i.e., if and only if E admits an atlas of local trivializations whose transition functions take values in $SO(r)$.*

Example 4.7.2 Every trivial bundle is orientable.

Example 4.7.3 Any almost complex manifold (M, J) is orientable. Every tangent space of M has an obvious structure of a complex vector space and we equip it with its canonical orientation. The tangent bundle of M becomes a smooth complex vector bundle and if $s_1, \ldots, s_r$ is its frame, then $s_1, \ldots, s_r, Js_1, \ldots, Js_r$ is an oriented frame of TM.

Example 4.7.4 The Möbius strip is not orientable. This can be seen by means of Problem 4.7.3.

Example 4.7.5 By the preceding considerations, a smooth manifold of dimension n is orientable if and only if it admits a non-vanishing n-form. The sphere S^n possesses such a form; it is defined by $\omega_a(u_1, \ldots, u_n) = (dx_1 \wedge \ldots \wedge dx_{n+1})(a, u_1, \ldots, u_n)$ for $a \in S^n$ and $u_1, \ldots, u_n \in T_a S^n$, $x_1, \ldots, x_{n+1}$ being the standard coordinates of $\mathbb{R}^{n+1}$. Thus, every sphere is orientable.

Example 4.7.6 The real projective space $\mathbb{RP}^n$ is orientable if and only if n is odd. More generally, the real Grasssmann manifold $Gr(k, \mathbb{R}^n)$ is orientable if and only if n is odd, see e.g., Fuks and Rokhlin [6], Chap. 3, § 2, Sect. 2, item 4.

Example 4.7.7 Let E be an oriented bundle. Consider every fibre E_x with the orientation opposite to the given one. In this way, we define an orientation of E: if $(s_1, s_2, \ldots, s_r)$ is an oriented frame of E. the frame $(-s_1, s_2, \ldots, s_r)$ yields the opposite orientation of the fibres of E.

Example 4.7.8 Let $E \to N$ be an oriented bundle and $f : M \to N$ a smooth map. Let $\widetilde{f} : f^*E \to E$ be the canonical map that is a fibre-isomorphism $(f^*E)_x \to E_{f(x)}$ for every $x \in M$. We use this isomorphism to transfer the orientation of $E_{f(x)}$ to the fibre $(f^*E)_x$ of the pull-back bundle. For every $p \in M$, there is a frame $(s_1, \ldots, s_r)$ of sections of E in a neighbourhood of $f(p)$ yielding the orientation of each fibre E_x. Then $(s_1 \circ f, \ldots, s_r \circ f)$ is a frame of the pull-back bundle f^*E such that the basis $(s_1 \circ f(x), \ldots, s_r \circ f(x))$ of every fibre $(f^*E)_x$ determines the orientation of $(f^*E)_x$. In this way, we define an orientation on the bundle f^*E.

Example 4.7.9 If E is an oriented bundle, endow every fibre E_x^* of the dual bundle E^* with the orientation induced by the orientation of the fibre E_x of E. This defines an orientation on the bundle E^*.

Example 4.7.10 The direct sum $E' \oplus E''$ of oriented bundles admits an orientation induced by the orientations of E' and E'' in an obvious way.

Problem 4.7.5 *Prove that every orientable line bundle is trivial.*

Problem 4.7.6 *Show that a rank r vector bundle E is orientable if and only if its determinant bundle $\Lambda^r E$ is orientable (hence trivial).*

Proposition 4.7.1 *Every vector bundle over a simply connected manifold is orientable.*

For a proof, we refer to Narasimhan [20], Chap. 2, § 2.7, Corollary 2.7.6. or Bott and Tu [2], Chap. 1, § 1, Proposition 11.5.

Suppose $E \to M$ is an oriented bundle and D is a connection on it. For any $x \in M$, there exists an oriented frame $(s_1, \ldots, s_r)$ of E in a neighbourhood of x. Take an oriented basis $(a_1, \ldots, a_r)$ of E_x. By Chap. 2, Corollary 2.9.1, there is a frame $e_1, \ldots, e_r$ in a neighbourhood of x such that $e_i(x) = a_i$ and $De_i|_x = 0$, $i = 1, \ldots, r$. The determinant of the transition matrix from the frame $(e_1, \ldots, e_r)$ to the frame $(s_1, \ldots, s_r)$ is positive at the point x, hence positive in a neighbourhood of x. Thus, the frame $(e_1, \ldots e_r)$ is positively oriented in this neighbourhood. By Chap. 2, Proposition 2.9.2, if D is a metric connection and $(a_1, \ldots, a_r)$ is an oriented orthonormal basis of E_x, we can find an oriented orthonormal frame $(e_1, \ldots, e_r)$ such that $e_i(x) = a_i$ and $De_i|_x = 0$.

4.8 The Euler Class of an Oriented Vector Bundle

The Euler class of an oriented bundle is defined via the Pfaffian of the curvature matrix of a metric connection on the bundle.

The Pfaffian of a $2k \times 2k$-matrix $A = [a_{ij}]$ is defined as

$$Pf(A) = \frac{1}{2^k k!} \sum_{\sigma} \epsilon(\sigma) a_{\sigma(1)\sigma(2)} \cdots a_{\sigma(2k-1)\sigma 2k}$$

where the sum is taken over all permutations σ of $\{1, \ldots, 2k\}$ and $\epsilon(\sigma)$ is the signature of σ.

Proposition 4.8.1 *Let A and B be real $2k \times 2k$-matrices. Then:*

(i) $Pf(B^t AB) = (\det B) Pf(A)$.

(ii) If $B \in SO(2k)$, then $Pf(B^{-1}AB) = Pf(A)$.

(iii) If the matrix A is skew-symmetric, $Pf(A)^2 = \det A$.

A proof of this proposition can be seen in Walschap [23], Chap. 6, Sect. 3, Proposition 3.1 and Corollary 3.1.

By Proposition 4.8.1 (ii), the Pfaffian is a $SO(2k)$-invariant polynomial function of degree k on the space $\mathcal{M}_{2k}$ of real $2k \times 2k$-matrices. More generally, let P be a $SO(2k)$-invariant polynomial function of degree p. Let $E \to M$ be an oriented vector bundle with a metric g. Denote by $\mathcal{A}(E, E)$ the bundle of skew-symmetric endomorphisms of E. Take a form φ on M with values in $\mathcal{A}(E, E)$. Let Φ_e be the matrix representing φ with respect to an oriented orthonormal frame $e = (e_1, \ldots, e_r)$ as in the considerations preceding Proposition 4.3.1. If $e' = (e'_1, \ldots, e'_r)$ is another oriented orthonormal frame, $\Phi_{e'} = T\Phi_e T^{-1}$ where T is the transition matrix from e to e'. This matrix is orthogonal and with positive determinant, so $T \in SO(2k)$. Hence, $P(\Phi^t_{e'}) = P(\Phi^t_e)$. Then setting $P(\varphi) = P(\Phi^t_e)$ locally, we obtain a globally defined differential form on M.

Remark 4.8.1 Because of the definitions of the curvature matrix and the curvature tensor in Chap. 2, to define the form $P(\varphi)$, we use here the transpose matrix Φ^t_e instead of Φ_e (see Example 4.8.2 below).

Proposition 4.8.2 *Let $E \to M$ be an oriented vector bundle of rank $r = 2k$ with a metric g. Let $R \in A^2(\mathcal{A}(E, E))$ be the curvature tensor of a metric connection D on E. Let P be a SO(2k)- invariant polynomial function on $\mathcal{M}_r$ of degree p. Then*

(1) The form $P(R)$ on M is closed.

(2) The cohomology class $[P(R)] \in H^{2p}_{dR}(M, \mathbb{R})$ does not depend on the choice of the metric g and the metric connection D on E.

Proof To prove claim (1), repeat the proof of formula (4.8), then the proof of Proposition 4.3.1, (1) making use of oriented orthonormal frames.

(2) Let $\widetilde{g}$ be a metric on E and $\widetilde{D}$ a connection on E compatible with this metric. Then for every $t \in [0,1]$, $g_t = (1-t)g + t\widetilde{g}$ is a metric on E and $D_t = (1-t)D + t\widetilde{D}$ is a connection compatible with g_t. Now, apply the arguments in the proof of Proposition 4.3.1, (2). □

Definition 4.8.1 Let $E \to M$ be an oriented vector bundle of rank r. If $r = 2k$ is an even number, take a metric g and a metric connection D on E. Let $R \in A^2(\mathcal{A}(E,E))$ be the curvature of D. The cohomology class

$$e(E) = \frac{1}{(2\pi)^k}[Pf(R)] \in H^{2k}(M, \mathbb{R})$$

does not depend on the choice of g and D, and is called the **Euler class** of E. The differential form $\frac{1}{(2\pi)^k} Pf(R)$ representing this class is called the **Euler form** (of (g, D)).

If the rank of E is odd, we set $e(E) = 0$.

Remark 4.8.2 The reason for setting $e(E) = 0$ for an odd rank bundle E comes from the definition of the Euler class in algebraic topology as an element of $H^r(M, \mathbb{Z})$.

Proposition 4.8.1 (iii) immediately implies

Proposition 4.8.3 *If E is an oriented bundle of even rank $r = 2k$, $e(E)^2 = p_k(E)$.*

Proposition 4.8.4 *If an oriented bundle E admits a nowhere vanishing section ξ, the Euler class of E vanishes.*

Proof Take a metric g on E and set $s = \xi/||\xi||_g$. By Chap. 2, Problem 2.4.4, there is a metric connection D on E such that $Ds = 0$. Let $\mathbb{R}s$ be the line subbundle of E determined by s and let $F = (\mathbb{R}s)^\perp$ be the subbundle orthogonal to it. Take an orthonormal frame $s_2, \ldots, s_r$ of F such that the orthonormal frame $s_1 = s, s_2, \ldots, s_r$ yields the orientation of E. If Θ is the curvature matrix of D with respect to this frame, $\Theta_{j1} = -\Theta_{1j} = 0$. It follows $Pf(\Theta^t) = 0$, hence $e(E) = 0$. □

Proposition 4.8.5 *(1) Let $f : M \to N$ be a smooth map, $E \to N$ an oriented vector bundle, and $f^*E \to M$ its pull-back bundle. Then*

$$e(f^*E) = f^*e(E).$$

(2) If E is an oriented bundle and $\overline{E}$ is the real bundle E considered with the opposite orientation, then

$$e(\overline{E}) = -e(E).$$

(3) If $E' \to M$ and $E'' \to M$ are oriented vector bundles,

$$e(E' \oplus E'') = e(E').e(E''),$$

where the product is taken in the ring $H^*_{dR}(M, \mathbb{R})$ of de Rham cohomology.

Proof Take a metric g and a metric connection on E. Let Θ be the curvature matrix of D with respect to an oriented orthonormal frame $s = (s_1, s_2, \ldots, s_r)$ of E.

The connection $\widetilde{D} = f^*D$ on the bundle f^*E is compatible with the metric $\widetilde{g}$ obtained from g via the canonical fibre-isomorphism map $\widetilde{f} : f^*E \to E$. Its curvature matrix with respect to the frame $\widetilde{s} = (s_1 \circ f, s_2 \circ f, \ldots, s_r \circ f)$ is $\widetilde{\Theta} = f^*\Theta$, which implies (1).

Let $\overline{\Theta}$ be the curvature matrix of D with respect to the frame $\overline{s} = (-s_1, s_2, \ldots, s_r)$. The transition matrix T from the frame s to the frame $\overline{s}$ is orthogonal and $det T = -1$. Then, by Proposition 4.8.1 (i), $Pf(\overline{\Theta}^t) = (det T^{-1}) Pf(\Theta^t) = -Pf(\Theta^t)$. Hence, $e(\overline{E}) = -e(E)$.

For a proof of (3), see Walschap [23], Chap. 6, Sect. 4, Theorem 4.2. □

We refer to Walschap [23], Chap. 6, Sect. 9, Theorem 9.1 for the proof of the following claim.

Proposition 4.8.6 *Consider a complex bundle E of rank r as a real bundle endowed with the orientation defined by the complex structure as in Example 4.7.3. Then $e(E) = (-1)^{[r/2]} c_r(E)$.*

Remark 4.8.3 If we consider the bundle E with the orientation induced by the complex structure as in Remark 4.7.1, then clearly $e(E) = c_r(E)$ This identity for $e(E)$ also appears in the literature.

If M is a compact manifold, the de Rham cohomologies $H^k(M, \mathbb{R})$ are finite-dimensional vector spaces and the number

$$\chi(M) = \sum_{k=o}^{k} (-1)^k \dim H^k_{dR}(M, \mathbb{R})$$

is called the **Euler characteristic** or Euler-Poincaré characteristic.

Example 4.8.1 The Euler characteristic of the unit sphere S^n is $\chi(S^n) = 1 + (-1)^n.1 = 2$ if n is even and $\chi(S^n) = 0$ if n is odd.

The classical Poincaré-Hopf theorem provides an important interpretation of the Euler characteristic $\chi(M)$ in terms of the behaviour of any vector field on M with isolated zeroes around such zeros.

Let ξ be a vector field on an open subset U of $\mathbb{R}^n$ vanishing only at a point a of U. If S is a sphere in U around a, the degree of the map $x \in S \to \xi(x)/||\xi(x)|| \in S^{n-1}$ is called the **index of** ξ at the point a. This number does not depend on the choice of S since any two spheres in U around a are homothetic. Now, let X be a vector field with isolated zeroes on an oriented manifold M. If z is an isolated zero, take a neighbourhood W of z such that: (1) X has no zero in W except z; (2) there is a diffeomorphism $\varphi : W \to U$ of W onto an open subset U of $\mathbb{R}^n$ that preserves the orientation, i.e., the linear isomorphisms $\varphi_{*x} : T_x M \to T_{\varphi(x)}\mathbb{R}^n$ send oriented bases to oriented bases. Then the index of the vector field $\xi = \varphi_* \circ X \circ \varphi^{-1}$ on U at the point $a = \varphi(z)$ is called the index of X at z and is denoted by $ind_z X$. This number does not depend on the choice of W and φ.

The following theorem is due to Poincaré and Hopf.

Proposition 4.8.7 *Let M be a compact oriented manifold and X a vector field on M with finitely many zeroes. Then the sum of the indices of X at its zeroes is equal to the Euler characteristic of M:*

$$\sum_{z\, is\, zero\, of\, X} ind_z X = \chi(M)$$

A nice reference on this matter is the Milnor [16] book on differential topology.

Since the Euler characteristic of any sphere of even dimension is not zero, the Poincaré-Hopf theorem implies the following claim known as the **hairy ball theorem**.

Proposition 4.8.8 *Every vector field on an even-dimensional sphere has a zero.*

The next statement is known as the generalized Gauss-Bonnet theorem.

Proposition 4.8.9 *If M is a compact oriented even-dimensional manifold with Euler form e, then*

$$\int_M e = \chi(M).$$

We refer to Walschap [23], Chap. 6, Sect. 7, Theorem 7.2 for a proof.

Example 4.8.2 Let (M, g) be an oriented Riemmannian manifold of dimension 2. Take an oriented orthonormal frame (e_1, e_2) and denote by $\Theta = [\Theta_{ij}]$ the curvature matrix of the Levi-Civita connection with respect to the frame (e_1, e_2). Then $Pf(\Theta^t) = \Theta_{21}$. If R is the curvature tensor, $R(X, Y)e_2 = \Theta_{21}(X, Y)e_1 + \Theta_{22}(X, Y)e_2 = \Theta_{21}(X, Y)e_1$ for $X, Y \in TM$. Hence, $\Theta_{21}(X, Y) = g(R(X, Y)e_2,$

$e_1) = g(R(e_1, e_2)e_2, e_1)(e_1 \wedge e_2)(X, Y)$. Note that $K = g(R(e_1, e_2)e_2, e_1)$ is the Gausssian curvature and $e_1 \wedge e_2$ is the volume form ω (the element of area). Hence, the Euler form is $e = \frac{1}{2\pi} Pf(\Theta^t) = \frac{1}{2\pi} K\omega$ and Proposition 4.8.9 gives the classical Gauss-Bonnet formula $\int_M K\omega = 2\pi\chi(M)$.

Bibliography

1. R. Bott, J. Milnor, On the parallelizability of the spheres. Bull. Am. Math. Soc. **64**, 87–89 (1958)
2. R. Bott, K.W. Tu, *Differential Forms in Algebraic Topology. Graduate Texts in Mathematics*, vol. 82, (Springer, 1982)
3. F. Brickell, R.S. Clark, *Differentiable Manifolds* (Melbourne, An Introduction (Van Nostrand Reinhold Company London, New-York, Cincinnati, Toronto, 1970)
4. L. Conlon, *Differentiable Manifolds* (Modern Birkhüser Classics (Birkhaüder, Boston, Basel, Berlin, 2001)
5. J.-P. Demailly, *Complex Analytic and Differential Geometry*, Université de Grenoble I, Institut Fourier, UMR 5582 du CNRS (2012), https://www-fourier.univ-grenoble-alpes.fr//~demailly/manuscripts/agbook.pdf
6. D.B. Fuks, V.A. Rokhlin, *Beginner's Course on Topology: Geometric Chapters* (Universitext, Springer, 1984)
7. C. Godbillon, *Géometrie Differentielle et Mécanique Analytique* (Collection Méthods (Hermann, Paris, 1969)
8. A. Gray, M. Barros, A.M. Naveira, L. Vanhecke, The Chern numbers of holomorphic vector bundles and formally holomorphic connections of complex vector bundles over almost-complex manifolds. Journal für die reine und angewandte Mathematik **314**, 84–98 (1980)
9. P. Griffiths, J. Harris, *Principle of Algebraic Geometry* (John Wiley & Sons, New York, 1978)
10. S. Helgason, *Differential Geometry, Lie Groups, and Symmetric Spaces. Pure and Applied Mathematics*, vol. 80, (Academic Press Inc., 1978)
11. M.W. Hirsch, *Differential Topology. Graduate Texts in Mathematics*, vol. 33, (Springer, 1976)
12. F. Hirzebruch, *Topological Methods in Algebraic Geometry* (Springer, 1978)
13. D. Husemoller, *Fibre Bundles. Graduate Texts in Mathematics*, vol. 20, (Springer, 1993)
14. S. Kobayashi, K. Nomizu, *Foundations of Differential Geometry*, vols. I. and II, (Wiley Classics Library. A Wiley-Interscience Publication. John Wiley & Sons, Inc., New York, 1963)
15. J.M. Lee, *Introduction to Smooth Manifolds, Graduate Texts in Mathematics*, vol. 218, (Springer, 2006)
16. J.W. Milnor, *Topology from the Differentiable Viewpoint* (Princeton University Press, Princeton NJ, 1997)

J. Davidov, *Vector Bundles and Connections*, Compact Textbooks in Mathematics,
https://doi.org/10.1007/978-3-032-07403-4

17. J.W. Milnor, J.D. Stasheff, *Characteristic Classes* (Princeton University Press, Princeton, NJ, 1974)
18. A. Moroianu, *Lectures on Kähler Deometry. London Mathematical Society Student Texts*, vol. 69, (Cambridge University Press, Cambridge, 2007)
19. J. Munkres, *Topology* (Prentice Hall Inc, Upper Saddle River, NJ, 2000)
20. R. Narasimhan, *Analysis on Real and Complex Manifolds. Advanced Studies in Pure Mathematics*, vol. 1, (Masson & Cie, Editeurs, Paris; North-Holland Publishing Co., Amsterdam-London; American Elsevier Publishing Co., Inc., New York, 1973)
21. B. O'Neill, *Semi-Riemannian Geometry: With Applications to Relativity. Pure and Applied Mathematics*, vol. 103, (Academic Press, Inc. New-York, 1983)
22. C. Voisin, *Hodge Theory and Complex Algebraic Geometry I. Cambridge Studies in Advanced Mathematics*, vol. 76, (Cambridge University Press, Cambridge, 2002)
23. G. Walschap, *Metric Structures in Differential Geometry. Graduate Texts in Mathematics*, vol. 224, (Springer, 2004)
24. F.W. Warner, *Foundations of Differentiable Manifolds and Lie Groups. Graduate Texts in Mathematics*, vol. 94, (Springer, 1983)
25. R. Wells, *Differential Analysis on Complex Manifolds. Graduate Texts in Mathematics*, vol. 65, (Springer, 2008)

Index

J. Davidov, *Vector Bundles and Connections*, Compact Textbooks in Mathematics,
https://doi.org/10.1007/978-3-032-07403-4

The manufacturer's authorised representative in the EU is Springer Nature Customer Service Centre GmbH, Europaplatz 3, 69115 Heidelberg, Germany. If you have any concerns regarding our products, please contact ProductSafety@springernature.com

Printed and bound by CPI Group (UK) Ltd, Croydon, CR0 4YY

07/07/2026

02160930-0002